The Ethics of Virtual and Augmented Reality

This book offers new ways of thinking about and assessing the impact of virtual reality on its users. It argues that we must go beyond traditional psychological concepts of VR "presence" to better understand the many varieties of virtual experiences.

The author provides compelling evidence that VR simulations are capable of producing "virtually real" experiences in people. He also provides a framework for understanding when and how simulations induce virtually real experiences. From these insights, the book shows that virtually real experiences are responsible for several unaddressed ethical issues in VR research and design. Experimental philosophers, moral psychologists, and institutional review boards must become sensitive to the ethical issues involved between designing "realistic" virtual dilemmas, for good data collection, and avoiding virtually real trauma. Ethicists and game designers must do more to ensure that their simulations don't inculcate harmful character traits. Virtually real experiences, the author claims, can make virtual relationships meaningful, productive, and conducive to welfare but they can also be used to systematically mislead and manipulate users about the nature of their experiences.

The Ethics of Virtual and Augmented Reality will appeal to philosophers working in applied ethics, philosophy of technology, and aesthetics, as well as researchers and students interested in game studies and game design.

Erick Jose Ramirez is Associate Professor of Philosophy at Santa Clara University, USA.

Routledge Research in Applied Ethics

For more information about this series, please visit: https://www.routledge.com/Routledge-Research-in-Applied-Ethics/book-series/RRAES

The Ethics of Virtual and Augmented Reality

Building Worlds

Erick Jose Ramirez

Routledge
Taylor & Francis Group

NEW YORK AND LONDON

First published 2022
by Routledge
605 Third Avenue, New York, NY 10158

and by Routledge
2 Park Square, Milton Park, Abingdon, Oxon OX14 4RN

Routledge is an imprint of the Taylor & Francis Group, an informa business

© 2022 Taylor & Francis

Library of Congress Cataloging-in-Publication Data
Names: Ramirez, Erick Jose, author.
Title: The ethics of virtual and augmented reality : building worlds / Erick Jose Ramirez.
Description: New York, NY : Routledge, 2022. |
Series: Routledge research in applied ethics | Includes bibliographical references and index.
Identifiers: LCCN 2021038403 (print) | LCCN 2021038404 (ebook) |
ISBN 9780367486266 (hardback) | ISBN 9781003042228 (ebook)
Subjects: LCSH: Virtual reality--Moral and ethical aspects. |
Augmented reality--Moral and ethical aspects. |
Virtual reality--Philosophy. | Reality--Philosophy.
Classification: LCC QA76.9.V5 R35 2022 (print) |
LCC QA76.9.V5 (ebook) | DDC 006.8--dc23
LC record available at https://lccn.loc.gov/2021038403
LC ebook record available at https://lccn.loc.gov/2021038404

ISBN: 978-0-367-48626-6 (hbk)
ISBN: 978-1-032-18147-9 (pbk)
ISBN: 978-1-003-04222-8 (ebk)

DOI: 10.4324/9781003042228

Typeset in Sabon
by Taylor & Francis Books

Contents

Illustrations

Figures

Table

Acknowledgements

This book is the product of many conversations with incredible people. It, quite literally, would not exist without them. If there are especially interesting sections in this book, it's almost certainly because of ideas exchanged with others. If there are sections of the book where arguments are less than good, that's probably a topic I should have spoken more with others about.

I want to thank my colleagues at Santa Clara University, especially Scott LaBarge and Larry Nelson, both of whom I've co-authored work with. Thanks also to Meilin Chinn for several wonderful conversations about action and perception. They too made their way into this book. I was extraordinarily lucky to have had Shannon Vallor (now at the Edinburgh Futures Institute) as a mentor when we were both at Santa Clara University. Thank you, Shannon, for being an excellent role model. I've also co-authored some work with my friend Per Milam. It was a stroke of amazing luck that Per and I ended up at UCSD for graduate school and I'm positive I'm getting the better of our partnership.

I've had the opportunity to meet and exchange ideas with people whose work I distinctly admire and would like to acknowledge here. Andrew Kissel arranged a fantastic panel session at the virtual 2021 Pacific Meeting of the APA that included myself, Kathryn Francis, and Melissa McDonald centered on VR, where I was exposed to a lot of useful ideas (and a new collaborator). Javiera Perez-Gomez, Jordan Wallace-Wolf, and Nick Harrison have all given me helpful feedback on my VR and AR work at various conferences. Javiera in particular has also co-organized the *Latinx Philosophy Conference* where I've shared my work on several occasions. I thank audiences there and elsewhere for engaging with the ideas in this book.

All of the VR simulations coming out of our experimental philosophy VR lab were created in collaboration with students at Santa Clara University including Miles Elliott, Carl Maggio, Mohit Gandhi, Lia Petronio, Dorian Clay, Kendall Moore, and Shelby Jennett. I'm consistently humbled by their ingenuity, talent, and philosophical acumen. Miles, Lia, Mohit, and I, in collaboration with Jocelyn Tan (another SCU alumna) have also fruitfully collaborated on written work, and I hope to continue working with them.

Funding and support for this project has come from several sources. Brian Green, Irina Raicu, and the rest of the folks at the Markkula Center for Applied Ethics have lent me intellectual and financial help throughout my career. Thanks to them for encouraging me to continue working on VR and AR ethics and for their friendship. My early work on VR would also not have been possible without grants from Oculus Education and the American Philosophical Association. It's been a pleasure working with Andrew Weckenmann and Allie Simmons at Routledge on this project.

Last, and most importantly, I want to thank my partner Maggie Levantovskaya and my parents Luis Ramirez and Leslie Delagneau for their love and support. Los quiero muchísimo.

1 Exploring Strange New Worlds

It's a warm, not-too-hot, summer day. A slight breeze wafts through the air, hitting my face. I'm writing this paragraph while lying on a chic outdoor sofa on the terrace of a mid-century home in the Hollywood Hills. Los Angeles sprawls below. From here, the views are perfect. If I look to my left, I see downtown, with all of its construction cranes. To my right, I see Century City. Even Santa Monica and the Pacific Ocean are visible, a pretty rare sight really considering the haze of the summer. Taking my eyes off this view for a second, I tap on a menu to move the 55-inch monitor I'm writing this paragraph on so that it floats more conveniently above my head. I shift my attention back to my work and write a few emails while listening to music.

Then I feel something touch my feet. It's the very specific bump that I've come to associate with my dog, Woody, poking my foot with his nose. He probably wants attention, or more food, or to be let out. I look to my feet and, for a moment, am surprised when I don't see the body I expected to see. I don't see any body, actually. I don't have any feet to be poked. Strangely, Woody isn't there either. Wrapped up in the hillside views and the flow of writing, I'd forgotten that I'm not really lying on a sofa in a multi-million dollar home in Los Angeles. I'm hours away in my decidedly more modest apartment in San Jose. It's also not the middle of a sunny day, it's 11:00 pm. The sun set hours ago. I'm suddenly struck by the fact that I've been wearing what's become a heavy HTC Vive virtual reality headset and that maybe I should leave my virtual Los Angeles terrace and get ready for bed![1]

For those of us who have had an experience like this, the immersive power of virtual reality worlds can seem fantastical. It's uncanny how virtual worlds can so easily, so naturally, and so quickly trick us into feeling like we're somewhere we really aren't. My Los Angeles experience was made possible because I was using a VR application called Bigscreen. Using Bigscreen, and other applications like it, I've been able to experience things that, even 15 years ago, I would have thought were forever outside my reach.

To name just a few of the things I've done recently: I've had a whale encounter on a shipwreck on the ocean floor. I've explored the Martian surface and heard the real sounds of the Martian summer wind. I've been a

DOI: 10.4324/9781003042228-1

navigator on a British bomber during the Berlin Blitz in the Second World War and seen the flashes of German antiaircraft artillery litter the skies around me. I've become a bit of coral and witnessed the damaging effects of ocean acidification and climate change. I've even experienced what it might be like to be kidnapped by a maniacal robot with seriously bad intentions. During the last academic term, I took an entire class of college students to a virtual movie theater where we saw films *about* virtual reality. During experiences like these, reality, and virtual reality, can start to blend together. Virtual experiences like these force us to ask interesting questions about what our lives, and values, might look like in the not-too-distant future and even ask questions about who we really are, how we choose to represent ourselves, and whether any of these ideas will last through the end of the 21st century. Virtual reality applications have been used to help train surgeons and football players, to treat patients with PTSD, to treat phobias of many kinds, and, at least according to some, to offer us the opportunity to experience what it might be like to be other people (and other animals) in ways that once seemed impossible. VR has enormous transformative potential, and the experiences I've described are only the tip of an interesting and, as we'll see, ethically fraught iceberg.

Until recently, experiences like the ones I'm describing were limited to a select few. Traditionally, virtual reality was locked away behind the walls of large, well-endowed university laboratories or government research facilities. The devices, software development experience, and computing power required to create successful virtual reality simulations were simply too large or expensive for commercial audiences to really take advantage of. Although the technologies we now know as "virtual reality" have been in development since at least the 1950s, the first true commercially available virtual reality devices didn't hit global markets until the early 1990s.[2] Even in these cases, technical limitations, and costs, combined to make it difficult for these early systems to find commercial success. The transformative potential of these early technologies ensured that development would continue.

Several advances needed to come together before virtual reality systems would coalesce into something able to deliver the experiences I've been able to share with students. These advances only recently happened and they've begun to change how we think about the range of possible experiences available to us. In 2016 a trio of commercial VR hardware platforms were released. First, Oculus VR (a division of Facebook Inc.) released the Oculus Rift. Later the same year, HTC (in cooperation with the Valve Corporation) released their own VR headset, the HTC Vive (this is the device that made my Hollywood Hills experience possible). Near the end of 2016, the Sony Corporation also released its own VR hardware, the Playstation VR, for use with its proprietary Playstation gaming platform.

The latest generations of the HTC Vive and the Oculus Rift, both released in 2019, represent, as of this writing, the current state of the art in commercial virtual reality systems. However, brewing in the background behind these large hardware developments, we've also experienced a quieter revolution in virtual reality technology. With the increasing sophistication and processing power of smartphones, smartphone-based VR systems have exploded in availability and popularity. Smartphones are currently the most widely accessible way that people are experiencing virtual and augmented realities. Most people who have experienced a virtual world have done so via smartphone VR.

While smartphone systems are less technically capable, they are ubiquitous, and with that ubiquity comes a kind of transformative power. In 2019, an estimated 130 million people around the world regularly had access to and used VR technology, and the VR marketplace is projected to grow to over $160 billion USD by 2023, partially fueled by the widespread availability of smartphone VR applications (Dujmovic 2019). We might be tempted to think that the gaming industry is behind all of this growth. However, VR development is not limited to companies developing games. In 2019 there was a 33 percent increase in VR development for educational simulations and a 27 percent increase in simulations aimed at workplace and skill training (Rubin 2019). In late 2019, Facebook Inc. announced that they would be releasing VR social media spaces which, one day, might supplant more traditional PC-based social media platforms (Grubb 2019). These technologies, and those under development, have placed us on the cusp of some very interesting transformations.

Although I've written here primarily about virtual reality, this is probably not the best way of talking about the entire ecosphere of products aiming to *augment* or *extend* our sense of reality. Commercial augmented reality hardware got off to an interesting start with the popularization, and later cancellation, of Google's "Glass" in 2012 (History 2012). I'll argue later that the distinction between augmented and virtual reality is an artificial one and that it's likely to disappear soon. What all of these technologies have in common is that they want to immerse us into experiences by adding simulated content (the distinction between augmented and virtual realities is really a way of signalling *how much* of that experience is made up of simulated content). More recently, in 2016 the Microsoft Corporation introduced the "Hololens," a "mixed-reality" tool which has found a degree of industrial success. Microsoft is currently working on a third generation version of the Hololens for commercial production. Augmented and virtual realities blend together to paint a picture of a not-too-distant future where a lot of our concepts about who we are, where we live and work, and what it means to have relationships will look different than they do today.

This book is about these technologies and their philosophical implications. As a philosopher trained in moral psychology, my aims, expertise, and interests lead me to focus less on the technical hardware discoveries that made these devices possible (though these are interesting and worthy of attention) and more on the ethical implications that these technologies will have for our sense of self, our relationships, and on our understanding of harm. Although science-fiction authors and futurists have been writing about technologies that sound like modern VR since at least the 1950s, philosophical discussions of virtual reality have been dominated by a thought experiment first introduced by a philosopher named Robert Nozick in his book, *Anarchy, State, and Utopia* (1974). There, Nozick introduced the concept of an "experience machine" that could recreate any experiences a user could wish for and which could make those experiences feel as real as our real-life experiences feel. Nozick, in part driven by a desire to show that hedonistic (i.e., pleasure-based) theories of the good were false, believed that nobody would (and that nobody *should*) use such a machine.

In the 45 years that followed, philosophers have raised issues not only about Nozick's assumptions about the value of experiences but also about the way he thought about his experience machine in the first place. It's time, many believe, to move beyond Nozick's understanding of experience machines and to consider virtual reality technologies as they're actually emerging. In doing so, we might end up showing when, and why, Nozick was wrong to think that virtual experiences always lack value. In this book, I assess the psychological underpinnings of how VR experiences can fool us (to trigger the feeling of "presence," in psychological parlance) but also the philosophical and ethical implications that these technologies introduce at the individual, institutional, and social levels. Because virtual reality devices like the HTC Vive, the Oculus Rift, the Playstation VR and AR technologies like Google's Glass and Microsoft's Hololens are so new, important technical, metaphysical, and ethical questions about them remain unanswered. In some cases, important ethical questions remain largely unasked as we experience what will likely be remembered as an early 21st century VR/AR gold rush.

I intend to answer some of these questions in this book. One set of questions I look at will deal with the psychological and ethical status of virtual experiences themselves. We might pause at this early point to ask what makes an experience virtual (the term might seem an oxymoron – what's virtual about the experiences I'm *actually* having?!). Throughout the book, I'll use the term "virtual experience" casually. That is, a virtual experience is an experience caused by virtual or augmented reality hardware. These experiences will share many features with other sorts of unreal experiences (dreaming, hallucinations, and so on) and I'll note these similarities and differences when they become relevant.

With that concept in place, we can ask a few questions about virtual experiences. For example, are virtual experiences different from experiences generated by other kinds of media? Are there interesting (philosophical or psychological) differences between reading a story, watching a story, and immersing ourselves in that same story by experiencing it as part of a virtual or augmented reality simulation? I'll argue that the answer is almost certainly yes. Virtual and augmented reality simulations can (though they don't always) offer a distinctly unique kind of experience that other forms of media cannot. I'll call these experiences "virtually real experiences" and their importance is central to the ethical issues I introduce and evaluate throughout the book.

Relatedly, there are questions we must ask about dealing with the effects of virtually real experiences. Can virtually real experiences harm the people experiencing them? Can virtually real experiences have value? If so, is that value different from our non-virtual, real-life, experiences? I'll argue that virtual experiences generally, and virtually real experiences in particular, are misunderstood by most psychologists and by some philosophers. I'll have to first show that virtual experiences can sometimes be "virtually real"; however, if I'm right about the existence of virtually real experiences then we'll see that they can be just as harmful (or just as valuable) as real-life experiences. It's not clear, in other words, that Nozick was right to so quickly conclude that the experiences people have in his experience machine lack value or that it would be wrong to want to have them. Many of the features that determine whether any given simulation produces virtually real experiences are also *under our control*. This fact should make us more cautious about how we think about the ethics of simulated worlds we design or play in. It will sometimes, I will argue, be wrong to create, and sometimes wrong to use, these kinds of virtual and augmented reality simulations.

There's another set of issues that VR and AR introduce. These issues revolve around our sense of self, the psychology of experience, and the nature of identity (what I'll call structural intersectionality later on). Currently, it's possible to download, in many cases freely, VR simulations designed to "expand" our moral horizons by giving us access to the inner lives of other people. The intention behind most of these simulations is to make us more empathetic by giving us a sense of what it's like to be someone very different from us. I'll spend a lot of time arguing that these simulations raise ethical issues of their own. For example, *1,000 Cut Journey* (Cogburn et al. 2018) is a VR simulation that claims to give its users insight into what it's like to experience anti-Black racism first-hand. Its creators claim that, in experiencing *1,000 Cut Journey*, "the viewer *becomes* Michael Sterling, a black man, encountering racism as a young child, adolescent, and young adult" (1,000 Cut Journey 2018). Some readers might already have qualms about the ability of a VR simulation to provide such an experience. According to its creators, even if you don't identify as male or Black, and even if you haven't experienced anti-Black racism first-hand, a simulation like *1,000 Cut Journey* can be designed to show you what it's like to be such a person, to experience such acts.

There are many simulations that claim to provide this sort of first-person, "in-their-shoes" sense of empathy. One simulation, called *Becoming Homeless* (Ogle, Asher, & Bailenson 2018), as the name suggests, claims that it gives its users a first-person experience of what it's like to lose your job and experience homelessness in a city very similar to San Francisco. Such experiences, assuming they're possible, can thus give us access to points-of-view otherwise inaccessible to us, and such access might even make us better, more empathic people if they can make us more understanding of why some people's lives are different and that include challenges different from our own. Some researchers have even argued that VR simulations can go further than this and give us some sense of what it might be like to be a non-human animal like a cow at a slaughterhouse (Ahn et al. 2016). Simulations like these, and many others, raise interesting questions about VR and AR's ability to change us (for good or for ill) and also force us to consider what metaphysical limits VR and AR technology can have. What is it like to be us, and can VR or AR simulations show us what it's like to be someone or something else?

But it's important that we don't take these simulations at face value. *Can* virtual reality simulations like *Becoming Homeless* or *1,000 Cut Journey* live up to their promises to make us more empathetic, more moral, people by showing us what it's like to be on the receiving end of injustice? I'll argue that in order to answer these questions we'll need to first have a better understanding of what empathy itself is so that we can make sense of what needs to be true for us to have an experience of "what it's like" to be anything at all – including yourself! Empathy, it will turn out, is an imprecise concept that names at least five neurologically and psychologically distinct capacities, and not all of these capacities are relevant to questions about knowing what it's like to be someone (or something). Once we have an understanding of empathy in hand, we'll see why VR and AR simulations, regardless of their cinematic ingenuity, can *never* give us access to the first-personal experiences of others.

Virtual and augmented reality can't show us what it's like to be a bat, or a cow, but it also can't show us what it's like to be Michael Sterling (that is, of course, unless we *are* Michael Sterling or very much like him).[3] This means, I'll argue, that we're often wrong about the content of our virtual experiences and we can thus be led by such simulations to make inferential, even ethical, mistakes because of these errors. Given that developers have a great degree of control over the content and structure of their simulations, I'll argue that it can be wrong to design certain sorts of VR experiences which, intentionally or unintentionally, are likely to mislead people about the nature of their experiences.

However, this doesn't mean that VR and AR can't give us new experiences and it doesn't mean that they can't help us become more sympathetic people. While these technologies can't show us what it's like to be someone else, they *can* give us a sense of what it would be like *for us* to be subject to at least some kinds of racial, economic, gendered, or ableist injustices. In doing so, I'll argue that we aren't getting a better sense of what these experiences are like for others, but that through these experiences we might learn important moral lessons about ourselves and our own values and beliefs which can make us more sympathetic to others. There is a sense of real moral progress that is possible with VR and AR that really does seem to be unique to these kinds of simulated experiences, but it lies in harnessing our capacity for sympathy, not empathy.

The unique possibilities afforded by virtually real experiences are, arguably, the most exciting feature of VR and AR technologies, and they've generated widespread interest. Over the last decade, psychologists and philosophers have begun to adopt VR as a new tool for exploring the mechanisms that ground our sense of right and wrong and for exploring our moral concepts. Psychologists are beginning to rid themselves of older pen-and-paper moral vignettes that once dominated the study of moral judgment, and replacing them with VR simulations of morally fraught situations.

Philosophers, long accustomed to using thought experiments as philosophical tools, are themselves exploring the advantages of creating simulated experiences in place of the imagination. Where before, subjects would be asked to imagine themselves facing a runaway trolley, they can now be placed in a virtual world with a virtual trolley and with virtual lives on the line. What should we think about this shift in approach? I'll examine a set of related questions about the nature of experience and experimental ethics connected to these methodological changes. Are virtual and augmented reality tools really better for exploring questions about our moral concepts or moral judgments? Do they offer a distinctive advantage to philosophers or psychologists over existing methods? Should experimental philosophers ditch thought experiments in their research and replace them with VR or AR simulations? What, if any, new ethical issues might these technologies bring with them into the laboratory or the classroom?

I'll argue that some of the excitement surrounding VR and AR research is misplaced. In far too many cases researchers fail to take into account the unique design challenges posed by virtual reality simulations and thus produce data that is often less useful than more traditional methods. In failing to take these challenges into account, they produce simulated worlds that just don't do the job they're being asked to do. Simulations designed in those ways will generate bad data (if used in a lab) or have poor learning outcomes (if used in the classroom). However, I'll argue that we have real reasons to believe that, when the specific design challenges of virtual worlds are met, VR and AR simulations can improve moral psychological research by generating more ecologically valid environments, and that in the classroom they're likely to have some advantages over the Thought Experiment Paradigm.

However, I'll also argue that using VR and AR in these ways can introduce serious ethical dilemmas. Much of the research to date has focused on VR specifically, though the concerns I discuss transfer to AR simulations of the same sort. If VR simulations can be designed so that we react realistically in them and so thus generate virtually real experiences, as I think they can, then we must rethink the ethics of using such simulations on people. It is possible to traumatize people using VR simulations in many of the same ways that people can be traumatized in real-life experiments. However, such simulations might be necessary if they're to be an improvement over pen-and-paper thought experiments.

Psychologists, philosophers, clinical ethicists, and even game designers should think very carefully about how they design their simulations to consider this very real, though virtually induced, kind of trauma. Is there a way of squaring this circle? That is, is it possible to design an ethically acceptable, ecologically valid, simulation of the trolley problem? What about ticking time-bomb cases? Are there any limits about what it is possible to simulate (technically or ethically)? I'll suggest a set of design guidelines for the construction of virtual worlds that are meant to address these questions. These guidelines should be helpful for both research and commercial developers of VR and AR content.

Putting these issues to one side, we can also ask questions about what our virtual behaviors say about our real selves. If we do something immoral in VR or AR, does that make us worse people, somehow, in real life? Should we be more careful about what we do in VR and AR simulations lest it impact how we think or act in the real-world? In some cases, the answer to both questions is yes! How should we think about our virtual actions and whether they reflect on our actual characters? Are our virtual friends *merely* virtual or can they take the place of more traditional friendships? Is it possible to flourish if our experiences are largely virtual?

Virtual and augmented realities have the potential to be profoundly transformative new tools with the capacity to shape how we see the world, ourselves, and our relationships. These transformations are not risk free, and ethical risks abound. Some of these risks are not unique to virtual and augmented realities, but others, I'll argue, are not only unique – they've gone largely unnoticed and under-appreciated. If we want to build ethical virtual worlds, then we'll need to be much more mindful of the relationship between how we construct our simulations and the purposes to which we put them.

In the rest of this chapter, I discuss the major questions and themes that I take up throughout the book and preview the main claims I defend. The second chapter focuses on empathy. As a generic term, to empathize with someone is to engage in at least one of five different kinds of activities with respect to the target of our empathy. I'll explain why it might be less confusing to do away with the generic concept of empathy and replace it with terms like empathic contagion, descriptive or normative mind-reading, in-their-shoes experiences, and "what-it's-like" experiences. I'll use these distinctions to critique a long-standing methodological practice among philosophers and psychologists: the perspectival thought experiment. These experiments, represented most popularly by the classic trolley problem, ask us to imagine that we are in some sort of moral dilemma and then ask us to answer that dilemma as if we were really facing it. Such thought experiments are impossible to actually carry out using the imagination, or so I'll argue, because of the psychological limits of empathy and the imagination. I close Chapter 2 by arguing that this is one place where VR and AR can help us make genuine philosophical and psychological progress. Although we can't actually imagine what it would be like to be in a ticking time-bomb torture case, a well-designed VR or AR simulation of these cases might actually give us a real sense of how we'd react, and give psychologists and philosophers better (more accurate) information about our real-life moral judgments.

In Chapter 3, I begin by laying out what I think are the elements of virtual experience, and explain how a virtual experience of an event can be different from imagining the event or reading about the event or watching a film about the event. In part, my claims will focus on the immersive and interactive nature of virtual experiences. Although all sorts of experiences can engage us, VR and AR experiences are much more likely to engage us in ways that look, and feel, like real-life experiences.[4] I'll argue that there are three pieces to this puzzle: perspectival fidelity, context-realism, and psychological features unique to each user. The first two pieces, perspectival fidelity and context-realism, refer to elements of a simulation's design and its content.

A flight simulator, especially one designed to train people to actually fly a plane, is almost certainly going to be high on perspectival fidelity and context-real in terms of its design and content. A sci-fi third-person shooter, on the other hand, is going to be a lot less likely to be context-real or perspectivally faithful. There's a wide gulf between simulations of the latter and former kind and while some of these differences are about the content of the simulation (flying vs. sci-fi destruction), the very structure of how these simulations represent their content is also different. These differences, as we'll see, are crucially important if we want to make sense not only of how to design good VR and AR experiments (or good classroom simulations or good games) but also to understand when and why VR and AR simulations become unethical. When a user is high on the dissociative spectrum, something I'll say more about later, and when a simulation is high in context-realism and perspectival fidelity, then such users are much more likely to have virtually real experiences in that simulation. Virtually real experiences, then, are those that are treated as-if they were real by the person experiencing them. Although research on virtual experience is a relatively young field, I'll suggest that we have good reason to believe that my analysis of the elements of a virtually real experience is on the right track.

I will then turn my attention to assessing the ethical consequences of these concepts in three different domains: experimental, philosophical, and social. Chapter 4 focuses on experimental ethics. There I explain the current debate about whether VR simulations provide researchers with more ecologically valid experimental options than traditional methods. Unsurprisingly, I'll argue that they often can, but only when such simulations are designed with perspectival fidelity and context-realism in mind. Experiments designed without these concepts in mind, of which there are many, are not only less likely to result in ecologically valid experimental environments, they're less likely to give researchers any really useful information about how their subjects really think about moral dilemmas. However, here too, ethical issues must be given more attention than they're currently being given. Using several case studies of experiments already conducted, I'll argue that Institutional Review Boards (committees which must approve of the use of human subjects for experiments) should be much more cautious about approving VR and AR experiments designed (intentionally or accidentally) to cause virtually real experiences. For all the same reasons that some real-life experiments are unethical to conduct, I'll argue that some VR experiments are also unethical and that unethical research is already being conducted that fails to heed this message.

Chapter 5 revisits the concept of empathy and its connection with virtual reality. This time, I'll look at the ethics of VR and AR empathy enhancement simulations. As of the time this book went to press, only VR empathy enhancement simulations have been widely explored, though the concerns I introduce in this chapter apply equally to both VR and AR empathy enhancing simulations. These VR simulations include the already-mentioned *1,000 Cut Journey* but also a whole host of others claiming to provide their users with some sense of what it's like to have conditions like autism, to be wheelchair bound, to witness coral bleaching as a coral living on a reef, and so on. Such simulations are typically used to engage in what philosophers call "nudging." Nudging is a term of art that is used to refer to intentionally designed features of an environment that impact, but do not determine, a person's choices. To be nudged is to be influenced, but not coerced. The layout of an environment can nudge you to take one path instead of another, a sale can nudge you to buy a product, etc. VR and AR simulations can nudge us if they affect (but do not determine) our choices, our behaviors, or our values.

Empathy simulations like the ones mentioned above are intentionally built as vehicles for nudging their users. To design a simulation like *Becoming Homeless* is to design a simulation that is intended to help nudge someone to think about homelessness differently. Building upon the arguments from the third chapter, I'll argue that it's unethical to create or use simulations like these for at least two reasons. First, these simulations mislead their users. They give them the impression of having access to the minds of other people (and in some cases the minds of non-human animals) when, in fact, they cannot. However, problems don't end with the misleading nature of these simulations, they're also manipulative. When we change our beliefs, values, policy positions, or relationships on the basis of experiences we had while using these simulations, our autonomy is diminished. This produces something I'll refer to as "The Intersectional Dilemma" for empathy enhancement. I'll appeal to four common moral frameworks to suggest that *all four* would agree that VR and AR empathy simulations of this kind are wrong to develop and wrong to use.

The sixth chapter aims to summarize the prior chapters of the book into a general code of ethics for virtual and augmented realities. In particular, I pay special attention to existing codes of ethics developed by the two largest global organizations for professionals working within the technology sector, the IEEE (The Institute of Electrical and Electronics Engineers) and the ACM (Association for Computing Machinery). I crystalize the ethical issues raised throughout the book up to that point into concrete guidelines for anyone looking to develop virtual and augmented reality applications. We can't risk moving forward with these technologies without incorporating ethics into the design process from the beginning, and current codes are not tailored to the specific ethical risks introduced by AR and VR technologies.

In the final chapter, I focus on issues of identity, self, representation, justice, and control. More specifically, in this chapter I'll show that VR, and to an even greater extent AR, are poised to create several challenges to these concepts in the 21st century. Our concepts of who we are and how we identify and re-identify ourselves and others are fundamental to almost all human relationships. The relationship between our sense of self and our physical body has also been radically changing, especially in the early decades of the 21st century. Until very recently, we had only limited powers to alter our physical appearance (using makeup, cosmetic surgery, and the like). Already, AR applications exist that let users try on virtual makeup or see what they may look like after a surgical procedure. While these uses of AR seem like natural extensions of our current practices, they are not. In this chapter I'll suggest that AR and VR force us, hopefully in advance, to create a framework for a better understanding of who and what we are when our sense of self is no longer bound by biology.

In a world where AR visual overlays are ubiquitous, for example, who should have control over how we appear to others? Should individual users control how they appear to others? If so, what do we do about trolls? Should individuals get to control how others appear to them? If so, how can we prevent deep faked AR sexual harassment? Should corporations or states control how individuals are allowed to represent themselves? That too seems problematic. Additionally, we humans have spent our entire evolutionary history, up to this point, relying upon convenient heuristics to identify and re-identify ourselves and others. In a world where AR is ubiquitous, such heuristics lose meaning and purpose. How can we know whether we're speaking with the same person over time if their appearance (and voice) can *radically* change with each interaction? Here too I'll canvas some solutions, all of them problematic, and I'll consider a few potential upsides to all of this identity confusion.

To say that there is much work to be done in terms of the ethics and metaphysics of virtual worlds is an understatement. It's tempting to simply enjoy the beautiful views from the virtual Hollywood Hills; however, while experiences like these can seem innocuous (and in many cases they are!), there are serious pitfalls to rushing into the development of virtual worlds without an ethics *for them*. Given the choice, we should build ethical virtual worlds. We begin that journey in the next chapter.

Notes

1 The breeze, in my case, was not coming from the VR simulation itself but from a very real ceiling fan, which really added to the immersion of my Los Angeles experience.
2 I'm referring here to Sega Corporation's "Sega VR," released in 1993, and the Nintendo Corporation's "Virtual Boy," released in 1995. Sega's system, while innovative, was so heavy and expensive that only large amusement centers/arcades could afford them. The Virtual Boy, while in many ways resembling modern virtual reality systems, was also fraught with technical and design limitations. It could deliver only a monochromatic and limited VR experience while simultaneously being hard to code for and difficult to use.
3 It's important to be careful here. It really will depend on what that simulation is claiming to show us. If it's about what it might be like for a neurotypical person named Michael Sterling to walk down an empty street, then assuming that there isn't anything about the experience that makes it unique to Michael (what we'll call "base cases" in Chapter 2), then such a simulation may very well succeed. A quick survey of so-called "empathy enhancing" simulations will show us that this is not how such simulations work. The point of such simulations is to give their users access to the experiences of people very different from their own along many intersectional dimensions.
4 In some ways, AR will have an advantage over VR in terms of producing virtually real experiences. Although I've already said that the I think the distinction between AR and VR is one of degree and not of kind, the fact that most AR experiences will take advantage of allowing the world to represent itself and make use of real-life haptics for at least some simulated events makes it easier, I predict, to trigger virtually real experiences in AR. Because AR technologies are newer and even less well researched than VR technologies, this is only conjecture on my end at this point.

References

1,000 Cut Journey. 2018, April 20. "1,000 Cut Journey" launches at Tribeca Film Festival. Retrieved from https://brown.columbia.edu/1000-cut-journey-launches-at-tribeca-film-festival/

Ahn, S.J., Bostick, J., Ogle, E., Nowak, K., McGillicuddy, K., & Bailenson, J.N. 2016. Experiencing nature: Embodying animals in immersive virtual environments increases inclusion of nature in self and involvement with nature. *Journal of Computer-Mediated Communication*. doi:10.1111/jcc4.12173.

Bigscreen [computer software]. 2016. Bigscreen Inc. Available from https://store.steampowered.com/app/457550/Bigscreen_Beta/

Cogburn, C., Bailenson, J., Ogle, E., Tobin, A., & Nichols, T. 2018. *1,000 cut journey*. ACM SIGGRAPH 2018, Virtual, augmented, and mixed reality, Article no. 1. Vancouver, British Columbia, August 12–16.

Dujmovic, J. 2019, July 15. Opinion: Here's why you will be hearing more about virtual reality. *MarketWatch*. Retrieved from https://www.marketwatch.com/story/heres-why-you-will-be-hearing-more-about-virtual-reality-2019-07-15

Grubb, J. 2019, September 25. Facebook Horizon is Oculus VR's Roblox-like social space. *Venture Beat*. Retrieved from https://venturebeat.com/2019/09/25/faceboo k-horizon-is-oculus-vrs-roblox-like-social-space/

History. 2012. The history of Google Glass. *Glass Almanac*. Retrieved from http://glassalmanac.com/history-google-glass/

Iñárritu, A.G. 2017. *Carney y arena (virtually present, physically invisible)*. United States: Fondazione Prada, Legendary Entertainment.

Luck, M. 2009. The gamer's dilemma: An analysis of the arguments for the moral distinction between virtual murder and virtual paedophilia. *Ethics and Information Technology*, 11, 31–36.

Nozick, R. 1974. *Anarchy, state, and utopia*. New York: Basic Books.

Ogle, E., Asher, T., & Bailenson, J. 2018. *Becoming homeless: A human experience*. Virtual Human Interaction Laboratory. Retrieved from http://vhil.stanford.edu/becominghomeless/

Pertaub, D.P., Slater, M., & Barker, C. 2002. An experiment on public speaking anxiety in response to three different types of virtual audience. *Presence Teleoperators and Virtual Environments*, 11 (1), 68–78.

Rubin, P. 2019, August 8. Want to know the real future of AR/VR? Ask their devs. *Wired Magazine*. Retrieved from https://www.wired.com/story/future-ar-vr-survey/

Thomson, J. 1976. Killing, letting die, and the trolley problem. *The Monist*, 59, 204–217.

Thomson, J. 1985. The trolley problem. *Yale Law Journal*, 94 (6), 1395–1415.

2 Imagination and the Limits of Empathy

Researchers interested in morality (including philosophers, psychologists, and neuroscientists) often ask odd questions. Here's an example:

> [s]uppose a fanatic, perfectly willing to die rather than collaborate in the thwarting of his own scheme, has set a hidden nuclear device to explode in the heart of Paris. There is no time to evacuate the innocent people or even the movable art treasures – the only hope of preventing tragedy is to torture the perpetrator, find the device, and deactivate it.
>
> (Shue 1978)[1]

... would you do it? Here's another example most of us, if only thanks to shows like *The Good Place*, have probably heard before: imagine that

> [a] trolley is moving towards five people on [one] track. You are standing at a switch. If you turn the switch, the trolley will be diverted to a side track, but there is one person on this side track. Turning the switch will result in that person's death, and the five people on the main track will be saved.
>
> (Bruers & Braeckman 2014)

Do you think you know what *you* would do in these situations? Odd questions indeed!

How might you go about answering questions like these? Should you imagine yourself at the scene? Should you picture the "fanatic" and what he looks and sounds like? To really answer these questions, should you try to imagine the sound of a ticking clock, counting down the seconds until the explosion in Paris? Is it important for you to think about specific art treasures at risk of destruction? Should you think about the people that might be killed? When faced with the trolley problem, should you imagine what the cries of the people stuck on the tracks sound like (notice here I'm assuming you're imagining that they're crying in the first place)? What about what the people on the track actually look like?[2] Should we even be imagining the scene at all or is the drama mere set dressing for what's really not meant to be a thought "experiment" at all?

DOI: 10.4324/9781003042228-2

When we do answer questions like these, should we try and imagine what we might *actually* do in the situations, as the questions seem to ask us to do, or should we take a more detached point of view and think instead about what we think we *ought* to do in the situation (even if we don't think we could bring ourselves to do it)? To add even more complication, should we instead think about what a morally better person than us would do if they were in our place? If we do these sorts of things, are we answering the original questions? Thought experiments raise these and other questions and it's important that we, and especially researchers interested in investigating moral judgments, figure out exactly how their own subjects are answering these questions when they use thought experiments to collect data.

The ticking-time-bomb thought experiment was first proposed by a philosopher named Henry Shue in 1978. Shue would probably not be happy with how his thought experiment tends to be portrayed today (i.e., as a potential justification of torture). In his original paper, Shue argued that, while it's not *always* wrong to torture someone, ticking-time-bomb scenarios are too unrealistic and idealized to prove this. When I've used the ticking-time-bomb experiment in the classroom, students tend to balk at the case entirely. Real life, they'll sometimes say, just doesn't match up to the world of the thought experiment and it can be hard for them to say what they really would do or even what a moral saint ought to do in such cases. They'll often want more information; sometimes they'll claim we have *too much* information (i.e., that we would never know all of these things in real life and so the situation is faulty). I tend to agree with students about these kinds of concerns.

This chapter takes aim at those who use thought experiments both as teaching tools and as a way of doing philosophical and psychological research. I set out to show why thought experiments like Shue's, while often rhetorically powerful and pedagogically useful, can make for bad moral psychology. Since at least the 1970s, moral psychologists have been caught in the grip of what I'll call the Thought Experiment Paradigm.[3] This paradigm rests on the intuition that short moral vignettes, like the examples above, can give us genuine insight into the structure of morality and into the psychology of real-life moral judgment. I'm deeply skeptical about much (but, like Shue, not all) of this research. In order to see why, I'll need to say a lot more about thought experiments but also about the psychological mechanisms that enable us to think about hypothetical cases. The central claim in this chapter is that thought experiments like Shue's often assume we have a capacity for a specific kind of empathic perspective-taking. Getting clearer not only about what empathy *is* but about whether we're capable of the kinds of empathy required to accurately answer ticking-time-bomb-style thought experiments will require some ground clearing. Ultimately, I'll end with deep skepticism of our ability to answer these sorts of questions using our imagination alone.

I begin the chapter by saying more about thought experiments. Thought experiments have a rich history in the academy. Though philosophers are probably most strongly associated with thought experiments, they are not alone: physicists, psychologists, and novelists have appealed to thought experiments for all kinds of argumentative, educational, and rhetorical uses. Although by the end of the chapter I'll argue that philosophers should abandon the use of a specific kind of thought experiment in their research, not all thought experiments are problematic and not all uses even of these problematic thought experiments will turn out to be illicit.

To clarify, I start by outlining one way of understanding what thought experiments are, how many different *types* of thought experiment there might be, and explain the specific kind of thought experiment I take issue with. I'll call these thought experiments *perspectival* thought experiments. Following this, I zero in on the concept of empathy to argue that empathy, as a scientific and philosophical concept, ambiguously refers to one of at least five different psychological capacities. Given this ambiguity, we're better off replacing the term empathy with direct references to these more specific empathic mechanisms. Perspectival thought experiments, I'll argue, assume that we have greater capacities for "in-their-shoes" and "simulation" empathies than we actually have and this finding will cause trouble for anyone who wants to use these thought experiments to learn more about moral concepts. In the end, perspectival thought experiments, of which Shue's is only one famous example, will almost always be impossible for us to carry out.

This seems bad. However, the upshot is that virtual reality and augmented reality technologies can come to a partial rescue. VR and AR simulations of thought experiments might actually allow us to successfully complete them. Due to the relatively early stage of AR development, VR simulations have received nearly all of the attention from psychologists and philosophers and I'll follow suit in this chapter by focusing on VR (though this landscape will hopefully change quickly). VR and AR can, in a very specific way, be tools that allow for genuine philosophical progress when it comes to questions about how we would actually behave in perspectival thought experiments. In the next chapter, I'll take a closer look at how VR and AR can actually do this and how we can design ethical and immersive VR and AR worlds. For now, let's take a closer look at thought experiments themselves.

Thought Experiments

Thought experiments are central to the way that philosophers and psychologists carry out their work. Given their central place in philosophical method, it's not surprising that philosophers use thought experiments and, as a result, that they've also devoted a good deal of time trying to figure out how thought experiments work. Many psychologists (Haidt & Bjorklund 2007) and neuroscientists (Greene et al. 2001) have joined philosophers in accepting the intuition that thought experiments, especially thought experiments of the kind I opened the chapter with, are good tools for probing the nature of moral judgment and moral cognition. All three fields have many adherents to this paradigm, the Thought Experiment Paradigm, and we can define it in terms of a commitment to these assumptions.

Among philosophers, conversations about thought experiments have historically focused on resolving questions about the *content* of a thought experiment and the outputs (judgments) produced by people imagining them. For example, when we think about the trolley, barreling down the tracks, and we arrive at some decision meant to address the situation, are we learning something *a priori* about the nature of ethics or are these thought experiments an especially colorful way of making garden-variety arguments (that could be made equally well without all of the imagery, flying sparks, and casualties) (Norton 2004; Brown 2011)? Similarly, philosophers have been interested in figuring out whether thought experiments might be more akin to mental models that, like scientific models, we manipulate to test intuitions (Gendler 2007; Meynell 2014).

Other philosophers have been less interested in questions like these. Instead, they examine what *kind* of evidence, or what kinds of reasons, thought experiments give us – assuming they give us reasons at all (Machery 2011; Allman & Woodward 2008; Bunzl 1996). It certainly seems as if philosophers and psychologists sometimes take the outputs of thought experiments as helping them to settle (or at least make progress on) longstanding philosophical debates (Di Nucci 2012; Liao et al. 2012; Huebner & Hauser 2011; Swann et al. 2010). It also seems that many believe that thought experiments give us evidence and insight into the nature of the moral psychological mechanisms we use to make everyday moral judgments (Haidt & Bjorklund 2007; Pastotter et al. 2013).

I intend to largely sidestep most of these debates here, interesting as they are. In part this is because my claims are agnostic about their resolution. It's not clear whether thought experiments are capable of settling philosophical debates or of what kind of evidence they can give us. My focus here is not on thought experiments in general, as a category, or their methodological or metaphilosophical virtues and vices. What is clear, I think, is that at least one specific kind of thought experiment *cannot* settle philosophical debates or reveal much about realistic moral psychology. They cannot do this because they're almost always impossible for us to complete successfully.[4]

The specific type of thought experiments I have in mind are identified in part by the functional role that they play in an argument. They are assumed, by those who use them, to play a reason-giving, sometimes called a probative, role. This assumption can be seen by those using thought experiments in both moral (Buckwalter & Turri 2016) and non-moral cases (Myers-Schulz & Schwitzgebel 2013). However, what's especially important about these thought experiments is *how* they're thought to give us reasons. Perspectival thought experiments are most easily identified by the *task* that they ask subjects to do. I call these kinds of thought experiments "perspectival" to make clear that they share an essential feature: in order to successfully complete a perspectival thought experiment, a subject has to shift their perspective from whatever their actual viewpoint happens to be into a new imagined situation and only then should they make a judgment (or note their beliefs, feelings, etc.) from their new, imagined, point of view.

Think back to the traditional trolley problem posed by philosopher Philippa Foot (1978). What would *you* do in that classic 1 vs. 5 dilemma? Do you switch the tracks, saving five people who were going to die without your intervention? In the process, this would make you causally (and some think morally) responsible for the death of the other person who would have survived without your intervention. Perspectival thought experiments require that we answer this question by first recreating the situation in our imaginations. We have to mentally put ourselves *into* the trolley problem and imagine what it would be like for us to actually be confronted, in the first-person, by the dilemma it spells out. We must realistically become faced with the time-sensitive decision to save five by killing one, and then, from that position, we can *see* what we would do, how we would feel, or what we would, in that moment, believe is morally justifiable for us to do.[5]

Perspectival thought experiments, if they're meant to do the work that moral psychologists often assume they do (i.e., give us new data about moral judgment), require that we do more than *predict* how we would behave. As we'll see, if they ask us to do that, then we're predictably terrible at these sorts of predictions. Instead, they require us to imaginatively engage in actual (imagined) moral problem-solving. Such experiments assume that we, or at least that most of us, are capable of specific feats of imaginary perspective-shifting.[6] Before I say more about why I think we *don't* have these capacities, I want to more clearly talk about the kinds of thought experiments that are problematically perspectival from those that are merely incidentally perspectival. In the end, my argument against perspectival thought experiments only affects the former.

I've been defining the perspectival thought experiment in part by the task that it requires of its subjects but also in terms of the role that such experiments play in the context of philosophical or psychological research. Both of these elements will matter when we assess when, and why, perspectival thought experiments lead us astray by assuming impossible capacities. Bryce Huebner and Marc Hauser (2011) have argued that there are at least two reasons why the results of perspectival thought experiments should matter to philosophers and psychologists:

> First, it is genuinely interesting to know whether there is a difference between folk-moral judgments about difficult cases and more tutored philosophical intuitions about these cases. Where there are such differences, there is a further question about which of these intuitions (if any) ought to be taken seriously in the construction of an adequate moral theory, and how tutoring affects our native intuitions ... Second, there

is a question about how different factors are evaluated from the stand-
point of folk morality … perhaps the folk-moral judgment that it is
permissible to turn the trolley results from being overly impressed by
the numbers where these have been made most salient. This is where
empirically driven moral psychology can make a crisp connection with
[moral] intuitions.

(80–81)

Hauser and Huebner are not alone in thinking that the Thought Experiment
Paradigm can connect our intuitions about perspectival thought experiments
to real-life moral mechanisms. Matthew Liao and his colleagues, for example, have argued that

[i]t is worth mentioning that the cases we have presented are just the
kinds of cases that many philosophers … have regarded as appropriate
for eliciting intuitions and have fueled much of the philosophical debate
about trolleys …

(Liao et al. 2012, 664)

As I understand them, moral psychologists like Huebner & Hauser and Liao
et al. believe that our judgments of perspectival thought experiments like the
trolley problem provide us with real, new, data about moral psychological
mechanisms. They also must think that this data is generated by the same
psychological mechanisms our real-life moral judgments use (or else they
wouldn't help us settle philosophical debates about morality in the first
place). They must also believe, given these commitments, that our judgments
about perspectival cases are sufficiently like real-life judgments to make safe
inferences from them about how real-world moral psychology operates. In
doing this, perspectival thought experiments provide evidence meant to help
settle (or at least make progress on) normative philosophical debates.[7] To
think along these lines is to say that perspectival thought experiments play a
probative role and it is the probative role that I want to ultimately deny
them. The probative role is not the only role that thought experiments can
play, however.

Thought experiments are often used to play a rhetorical role in a discussion, functioning less as an argument itself and more as a tool for non-
rational persuasion. This use is rare among philosophers though perhaps
more common in political and social life (one familiar rhetorical perspectival
thought experiment: it's the middle of the night and a global emergency is
unfolding and a call reaches your head of government. Strong moral lea-
dership can mean the difference between calamity and salvation. Whom do
you want on the receiving end of that call?).[8]

In the classroom, thought experiments are less likely used for their per-
suasive or argumentative powers than for what Michael Davis (2012) has
called a "heuristic" function. A thought experiment serves a heuristic

function "if intended to help us recall, understand, or discover" something (Davis 2012, 3). To have a heuristic function, a thought experiment essentially makes a concept clearer or helps us better understand the consequences of particular theory. In such cases, thought experiments act like examples. They can make complex or abstract matters easier to understand. Albert Einstein's famous (1920) thought experiment about simultaneity judgments by observers on a train is clearly intended to play a heuristic role and not a mere rhetorical role:

> We suppose a very long train travelling along the rails with the constant velocity v and in the direction indicated ... People travelling in this train will with advantage use the train as a rigid reference-body (co-ordinate system); they regard all events in reference to the train. Then every event which takes place along the line also takes place at a particular point of the train ... Are two events ... which are simultaneous *with reference to the railway embankment* also simultaneous *relatively to the train?* We shall show directly that the answer must be in the negative.
>
> (25)

Thought experiments like Einstein's are not obviously intended as probative perspectival thought experiments. We don't need to imagine what our point of view as a passenger on the train would be in order to complete the experiment or understand its purpose. To the degree that we think we're able to imagine this, we might be engaging in perspective-taking; however, this is not essential to the function of Einstein's thought experiment. Given specific axioms about motion and of the speed of light, it will follow that the answer to whether someone will judge whether two events occur simultaneously will depend on the frame of reference the observer is using. This sort of thought experiment is intended to *clarify*, to bring out the implications of, a theory or principle. Thought experiments like these don't generate *new data* but instead point out the logical consequences of our commitments.

During a discussion in an introductory ethics course on act-utilitarianism, an instructor might use the trolley problem in this way. She might make clear to her students that, given certain assumptions about the values of lives, that the act-utilitarian would be committed to the view that switching the tracks was morally required (or not, depending on the specific circumstances). Immanuel Kant's shopkeeper examples from the *Groundwork* can be seen in a similar light (Kant 1785/1998). Kant used these experiments to help illustrate just what it is that the concept of the Goodwill encompasses. Good behavior, by itself, isn't enough to actually manifest a goodwill and we can see that by seeing what Kant says about the maxims that ground the different shopkeepers' behaviors toward their customers. There isn't anything obviously wrong with using thought experiments heuristically; they can be extremely useful educational tools. It's important to keep in mind that even perspectival thought experiments like the trolley problem, the ticking time bomb, or Kant's shopkeepers can have entirely acceptable heuristic uses.

I've argued that the most important features of the perspectival thought experiment include the following conditions:

1   In order to successfully complete them, subjects *must* imaginatively simulate the conditions of the thought experiment, they must then transport themselves into the relevant perspective, and from that imagined position, they are asked to make a judgment (or note their feelings)
2   They are understood to serve a probative role in an argument or
3   They generate new data for researchers interested in the nature of moral judgment or moral properties

As a final example of a perspectival thought experiment, consider a question that Huebner and Hauser (2011) posed to subjects during the course of one of their own experiments:

> You are standing near the railroad tracks and notice an empty boxcar coming down the tracks, moving fast enough to kill anyone that it hits. If you do nothing, the boxcar will continue along the main track, killing five people who are walking down the main track. There is a switch nearby that you can use to divert the boxcar onto either of two side tracks that split off from the main track in opposite directions. There is one person walking along the right-side track. So, if you flip the switch to the right, the boxcar will hit and kill this person. Your foot is stuck in the track on the left-side track. So if you flip the switch to the left, you will be hit and killed by the trolley. What should you do?
>
> (83)

It's hard to read this question in anything other than a perspectival sense. After all, it's *you* who is standing near the railroad tracks; it's *your foot* that you're to imagine is stuck on the track and it's *you*, in that situation, who must make a decision.

To correctly answer Huebner and Hauser's question, you must therefore imagine that all of the conditions in the experiment are actually happening to you. Assuming that we're able to perform this feat of imagination, you must then make a choice from within this imagined context. *That* is the choice that Huebner and Hauser are interested in recording for their experiment. Furthermore, they take their data to provide new reasons (i.e. that it has a probative function) that speak against philosophical positions:

> The data reported in the previous section minimally suggest that [Judith] Thomson's intuition is radically at odds with the commonsense

intuition of what a person should do when she is faced with a difficult moral trilemma such as the three-track bystander case. If Thomson were right, we would expect it to be far more transparent that it is immoral to turn the trolley, and this increased transparency should be reflected, at least to some extent, in the folk-moral judgments that are offered in response to this case.

(Huebner & Hauser 2011, 87)

This experiment is thus firmly set within the context of the Thought Experiment Paradigm. Perspectival thought experiments like these, because they require their subjects to place themselves into imagined simulations, are problematic. I haven't yet explained *why* perspectival thought experiments are problematic nor have I said why these problems should lead us to adopt a new VR/AR methodology in their place. To do that, I'll need to say more about how philosophers and psychologists understand empathy and per-spective-taking.

Empathy

Empathy is a word with many meanings. Colloquially, we often use the words empathy and sympathy interchangeably. In this section, I disentangle the capacities referred to by the name empathy so that we can be clearer about what it is that perspectival thought experiments presuppose we're capable of. I'll argue that, despite some seeming evidence to the contrary, we aren't really able to do what perspectival thought experiments ask us to do. When we say, for example, that we would switch the tracks in the trolley problem (or when we say that we would judge that it would be morally permissible or morally required for us to switch the tracks) we're system-atically deluded not only about the output of our imaginative exercise, but also about our ability to imagine the situation accurately in the first place.

To begin, it's useful to distinguish empathy from sympathy. Though any distinctions we make will be controversial, empathy and sympathy are best distinguished as the capacities for *feeling with* and the capacity for *feeling for* a target. To sympathize with someone is to feel for them but not to feel with them. To empathize is to do the reverse. We probably would prefer a sympathetic doctor over an empathetic one, for example. We want our doctors to care about us, to want us to be as healthy as we can be, and to perform their duties with that care and concern in mind. What we don't want is a strongly empathetic doctor. An empathetic doctor would have trouble carrying out some of their basic duties because their empathic responses are likely to get in the way of their being sympathetic to us. Ima-gine a doctor trying to set a broken leg while simultaneously feeling their patient's pain in their own leg! Imagine a surgeon feeling in their own body the scalpel cuts involved in the surgery they're performing on their patient! Of course, the capacity for "feeling with" can be described in many different ways, some of them compatible with being a good doctor. Let's break this down further.

Psychologist Paul Bloom has argued that "[t]o empathize with someone is to put yourself in her shoes, to feel her pain" (2014). This is an interesting definition as it combines two distinct capacities under the single banner of "empathy." Take the capacity to feel someone's pain. Although Bloom runs them together, it's important that we keep Bloom's two forms of empathy distinct from one another. Let's call the "feeling someone else's pain" form of empathy "mirroring" empathy. To empathize with someone in the mirroring sense is to, quite literally, feel someone else's pains and pleasures. To a certain degree, we're almost all able to engage in mirroring empathy and we're pretty familiar with these empathic experiences. We engage in mirroring empathy when we find ourselves reflexively smiling when someone smiles at us or when we wince in mirrored pain when we watch someone fall and hurt themselves. Because mirroring empathy is experienced as something we can catch from others, psychologists sometimes call this kind of empathy, "contagion" empathy (Goldman & Jordan 2013) (see Figure 2.1).

There are good reasons for distinguishing mirroring empathy from other kinds of empathy. First, mirroring empathy is, unlike the other forms of empathy I'll discuss, non-cognitive. What this means is that we don't have to consciously think about a target in order to empathize with them via mirroring.[9] Say, for example, that you watch someone take a bad fall near you. They're groaning and reaching for their badly skinned knee. In that moment, you might find yourself feeling mirrored pain in your own knee. The same goes for positive experiences. We tend to mirror the emotional expressions of those around us and feel similar feelings as those around us feel (McDonald & Messinger 2010). We do this without having to consciously work ourselves into

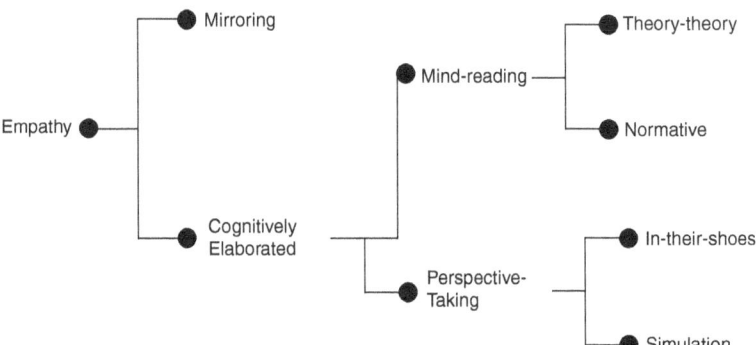

Figure 2.1 Varieties of empathy
Source: Adapted from: Ramirez, E. 2016, "Empathy and the limits of thought experiments." *Metaphilosophy, 48 (4)*, 504–526.

those feelings.[10] For most of us, this just happens on its own and we seem to "catch" the feelings of others.

Second, the neural correlates of mirroring empathy appear distinct from those of other (more cognitively elaborated) forms of empathy (Hobson 2002). This capacity is referred to as mirroring empathy precisely because of its functional connection to "mirror" neurons in the mammalian somatosensory cortex (Iacoboni 2009). These capacities are both functionally and neurologically distinct from the other forms of empathy and thus serve as (fairly strong) evidence of their being an independent empathic mechanism. Third, although not unique among other forms of empathy, the capacity for mirroring empathy is *dimensional*. [11] The capacity for mirroring empathy can, in other words, be thought of as a spectrum capacity whose boundaries are characterized, controversially, via pathology.[12]

At the extreme high end of the mirroring empathy scale are those with a condition diagnosed as "mirror-touch" synesthesia (MTS) (Ward & Banissy 2015). Individuals with MTS can find it difficult to leave their homes or watch television because their reactions to the physical and emotional pains and pleasures of others make life difficult. A person with MTS mirrors the emotions of others especially strongly relative to neurotypical people. These reactions can be so strong that individuals with MTS often claim to feel these feelings just as much, if not *more* so, than what those whom they empathize with actually feel. One individual with MTS has written that she has "never been able to understand how people can enjoy looking at bloodthirsty films, or laugh at the painful misfortunes of others when I can not only not look but also feel it" (Choi 2007). Giving their extraordinarily strong mirroring empathy responses, this kind of reaction makes sense.

At the extreme low end on the mirroring empathy scale are individuals with a condition known as psychopathy.[13] Psychopaths are characterizable in many ways, all of them controversial, but a very common feature among different diagnostic devices for psychopathy is a deficiency, even at a young age, in a capacity called "empathic distress" (Hobson 2002; Marsh et al. 2013). This capacity is also largely grounded in mirroring empathy systems. Empathic distress refers to our nearly automatic ability to feel pained by the pain of others and some philosophers believe that it's a necessary component of neurotypical moral development (Greenspan 2003; Nichols 2007; Ramirez 2019).

Contrast mirroring empathy with the other form of empathy that Paul Bloom mentioned: the capacity to put yourself in someone's shoes. This form of empathy is cognitive, *not* non-cognitive. It takes a lot of effort to imagine what it might be like to be in someone else's position, to be in her shoes, in a way that feeling pained by the pain of others doesn't. Referring back to Figure 2.1, we can see that the main branch in the empathy family is composed of these "cognitively elaborated" forms. Consider, for example, the differences between the following ways in which we might become aware of what it's like to be in someone's shoes.

David Hume (1738/2007) argued that "extensive" sympathy (a capacity that is much closer to what we're calling cognitively elaborated empathy) ought to allow us to *correct* our moral judgments in a way that removes emotional biases like proximity (favoring those closer to us than those distant from us) and self-interest (judging that situations that benefit us are morally better than those that don't):

> We blame equally a bad action, which we read of in history, with one perform'd in our neighborhood t'other day: The meaning of which is, that we know from reflexion, that the former action wou'd excite as strong sentiments of disapprobation as the latter, were it plac'd in the same position.
>
> (Hume 2007)

To correct our moral judgments in the way that Hume suggests requires that we engage in a very different kind of empathy. This form of empathy appears to require that we "reflect" on how some historically bad action (e.g., the assassination of Abraham Lincoln) *would* be judged by us if it were occuring today (or, alternatively, how we might feel if we were alive then). This requires that we be able to accurately imagine what it might be like for us to live in this alternative circumstance and note how we would feel from within it. Because it requires that *I* put myself into some other position (that is, I am not imagining that I am someone other than myself), let's call this form of empathy "in-their-shoes" empathizing.[14] I'll say more about in-their-shoes empathy in a moment.

Contrast in-their-shoes empathizing with a capacity that it's often confused with, the capacity for mindreading. Mindreading is perhaps an unfortunate name for this psychological capacity since it doesn't exactly mean what we might think it means. When philosophers and psychologists talk about mindreading, they're referring to a capacity (confusingly also sometimes called theory-theorizing), that most of us use to understand not only that other people's minds are different from our own (i.e., they need and want different things and have access to different information than we do), but also the capacity to make predictions about how other minds work. To think of empathy in a theory-theory/mindreading way means thinking of empathy as "the process by which one attains a cognitive grasp of, belief about, or knowledge of another's mental states" (Battaly 2011). Notice that, unlike in-their-shoes empathy, mindreading/theory-theory empathizing doesn't require that we imagine what things look like from some other point of view (our own or anyone else's). Theory-theory empathy doesn't actually require that we *feel* anything at all, it demands less of the imagination.

To successfully theory-theory empathize with someone requires that we possess a folk psychological theory that lets us understand other minds in a general way. Folk psychology is not psychology. Folk psychology, instead, refers to a store of knowledge most humans have that lets them explain and predict other people's behavior (Andrews 2008). Pairing folk psychology with specific knowledge about a person and using this knowledge to predict their behavior gets us to mindreading/theory-theory empathy. Suppose you know that your friend Sam is thinking about a career change. You've known Sam for years and know that she recently received a job offer that, while it pays less than she currently makes, is in a field that she's always wanted to work in. You also know that at this point in her life Sam would probably say that she wouldn't miss the extra money from her current job. What is Sam going to do?

If we successfully theory-theory empathize with Sam, then we'll predict that she'll take the new job. Notice that we do this *without* imagining what it would be like *to be* Sam or even what *you* might do if you were in Sam's place. Theory-theory empathy, while cognitively elaborated, is not a uniquely emotional or even an imaginative psychological process. As the name implies, this form of empathy requires the application of a theory to a set of data (what we know about a person) to generate a prediction about their thoughts, beliefs, feelings, or future behavior.

Theory-theory empathy is also a spectrum capacity. Individuals with autism spectrum disorder have often been characterized as having difficulty engaging in theory-theory empathy (Colle, Baron-Cohen, & Hill 2007).[15] Earlier, I noted that psychopaths often have difficulty with a form of mirroring called empathic distress. Interestingly, psychopaths do *not*, at least not as a matter of course, appear to have a corresponding problem with theory-theory empathy (Richell et al. 2003; Dolan & Fullam 2004). Elsewhere (Ramirez 2019), I've argued that data like these provide evidence for seeing these forms of empathy as distinct and as also one possible explanation for why psychopaths tend to be so successful at manipulating others (by understanding how other minds work) while simultaneously not feeling guilty about what they do (because they are not pained by the pain they cause others).

Some philosophers have argued that theory-theory empathy can also take on a normative character (Simmons 2014). In normative theory-theory empathy, "our concerns … may still count as empathy if we accept a broader sense of empathy which involves empathizing with how others ought to feel and not just how they actually feel" (Simmons 2014, 98). Instead of predicting what it is that Sam *will do* or *will feel* about her job offer, normative theory-theory empathizing is a way of predicting what Sam *should do* or how she *should feel* about the offer. We can pair folk psychological theory with details about Sam's own psychology and couple all of that information with our own preferred moral theory to argue that Sam *ought* to take the job (or not take the job) and that she *ought* to feel good about her decision (and maybe a little anxious) even if she *doesn't* do or feel any of these things. This might strike some readers as an odd way of talking about empathy. I agree.

I include normative theory-theory empathy here for completeness (and to show that empathy, as a term, can represent different psychological capacities). It seems odd to refer to normative theory-theorizing as an instance of empathizing, as philosophers like Aaron Simmons suggest. It's more natural, in my view, to call this sort of prediction a kind of applied moral philosophy. One way of thinking about applied ethics is as the branch of ethics where we appeal to moral theory to tell us what we (or others) ought to think or feel or do. It might even be more appropriate to call this kind behavior sympathy rather than empathy. To say that Sam ought to feel good even if she doesn't feel this way isn't to feel *with* her but, possibly, to feel *for* her (to wish her well or direct her feelings in a way that we believe would be good for her). Researchers working within the Thought Experiment Paradigm should not want subjects in their labs engaging in normative theory-theorizing when they respond to perspectival thought experiments.

Mirroring empathy, by definition, is a way of feeling with our target. If our target is in pain, we *too* feel pain when we empathize with them in a mirroring sense. Theory-theory empathy, descriptive or normative, doesn't require this kind of co-feeling. Other forms of cognitive empathizing do call for more active and imaginative engagement. Perspectival thought experiments, insofar as they are perspectival, don't seem to require mirroring empathy.[16] They also don't seem to require theory-theory empathizing. In order to respond to a perspectival thought experiment, what kind of empathic capacities must someone have? Let's consider another classic perspectival thought experiment often discussed by philosophers, psychologists, and neuroscientists:

> [Y]ou are standing on a footbridge over the trolley track. You can see a trolley hurtling down the track, out of control. You turn around to see where the trolley is headed, and there are five workmen on the track where it exits from under the footbridge. What to do? Being an expert on trolleys, you know of one certain way to stop an out-of-control trolley: Drop a really heavy weight in its path. But where to find one? It just so happens that standing next to you on the footbridge is a fat man, a really fat man. He is leaning over the railing, watching the trolley; all you have to do is to give him a little shove, and over the railing he will go, onto the track in the path of the trolley. Would it be permissible for you to do this?
>
> (Thomson 1985)

Let's call this version of the trolley problem the Bridge Problem. The Bridge Problem will re-emerge in later chapters when I argue that it's impossible to create a simulation (VR or AR) of thought experiments that share the same features that the Bridge Problem has.

As with other perspectival thought experiments, to address the Bridge Problem correctly (i.e., to respond to it *as* a perspectival thought experiment), we must engage in a unique kind of perspective shifting. It's unlikely (I hope!) that any of us are reading this paragraph while also standing on a bridge, facing a runaway trolley, and mulling over the moral permissibility of pushing someone onto its path. Because of that, we need to imaginatively simulate the content of the thought experiment to address its moral variables in the way the Bridge Problem calls for. What *would* we conclude is morally justified (or morally allowed, or morally forbidden) if we were on the bridge in that specific context?

Notice that normative theory-theorizing isn't what's called for here. Normative theory-theorizing wouldn't generate probative data (it wouldn't tell us anything we didn't already know about our moral commitments). The point of the Bridge Problem seems to be that we *learn* something about our moral intuitions or moral practices. Concluding that, because we're utilitarians, we must therefore push the man onto the path of the trolley, all things being equal, might be what our normative theory says we ought to do. However, the Bridge Problem is meant to create a space where we can generate new intuitions that might cause trouble for our pre-existing ethical frameworks and give us fodder for reflective equilibrium.

If, when actually faced with a situation like the Bridge Problem, I conclude it would be *morally impermissible* to push the man, then I might not be a consistent utilitarian after all. To answer the question posed by the Bridge Problem I need to know what I really would find justifiable in that situation so that I can test my own consistency. This means that perspectival thought experiments assume we're capable of creating perspectival mental simulations and of generating new data from the outputs of those simulations. This kind of empathy, as Figure 2.1 suggests, requires perspective-taking.

Perspective-taking empathy can take one of two forms. We've already discussed the first form of perspective-taking empathy: in-their-shoes empathy. To empathize with someone in the in-their-shoes sense requires that we shift our perspective but doesn't require that we be able to imaginatively change our identity. Return to the Bridge Problem. To answer Thomson's question, I need to put myself into the bridge situation to see what I would feel or think or to see what I really would conclude is justifiable to do. I don't, however, have to imagine what it would be like to be someone else facing the same situation.

The cognitive task of imagining what the Bridge Problem would look like for *me* to be in that situation is different (and arguably less demanding) than the task of imagining what it would be like *for someone else* to face it. I'll follow philosopher Alvin Goldman's lead in referring to the "what it's like to be someone else" form of empathy as "simulation" empathy. According to Goldman, this kind of empathy

> involves an attempt to replicate or reexperience the target's state via a constructive process. Exploiting prior information about the target, the mindreader uses enactment imagination to reproduce in his own mind what might have transpired, or may be transpiring, in the target.
>
> (2011, 38)

To see the difference, we can go back to Sam's job offer.

Suppose that Sam and I are very different kinds of people. She's much more of a risk-taker than I am and much braver in the face of danger than I would be. Now suppose that I'm trying to empathize with Sam in a situation where her life is in danger (say she's confronted by a mountain lion while out hiking alone). If I'm empathizing with her in the in-their-shoes sense, then my task is to imaginatively place *myself* into the same situation as Sam. To see, in other words, from behind my own eyes what I would do if I were to come face-to-face with a mountain lion while hiking alone. If I'm simulation empathizing with Sam, my task is different.

As I would with in-their-shoes empathizing, I must still imaginatively recreate the mountain lion scenario. However, instead of placing *myself* in that situation, I must also try to imagine what it would be like *to be Sam* while facing the mountain lion. To "replicate or reexperience the target's state," as Goldman put it. I must see the situation through Sam's eyes, not my own. Notice that the outputs of these two forms of empathy are likely to be very different given how different Sam and I are. If I successfully place myself into Sam's place (i.e., in-their-shoes empathize with her), I'm likely to be terrified. Mountain lions are scary and dangerous and I'm easily frightened. Running into one unexpectedly is probably going to be terrifying for me. However, Sam isn't like me. If I successfully empathize with her in the simulation sense, then I probably shouldn't feel afraid at all. Sam's character is different enough from mine that she might feel some anxiety, but not fear. She knows how to handle these situations, I don't. There are likely to be many ways in which Sam would see and think about the situation that would be different from how I would see and think about it. If I successfully imagine what it's like to be Sam, I'm going to think and feel differently.

Thinking back to the ticking-time-bomb or the trolley or the bridge, perspectival thought experiments aim to recruit either in-their-shoes- or simulation-based empathic mechanisms. It isn't possible to complete a perspectival thought experiment *without* recruiting one of them (exactly which will depend on the perspectival thought experiment). However, it really looks like simulation based empathy is impossible for us to ever successfully carry out, at least if we limit ourselves to what we can do with our imagination. Likewise, it's *almost always* impossible to in-their-shoes empathize with others. If I'm right about this, then that's bad news for the Thought Experiment Paradigm. However, if *I am* right about this, then there's a good possibility that we can supplant the Thought Experiment Paradigm with a new, virtual reality based, method of conducting perspectival thought experiments. So what's wrong with in-their-shoes- and simulation-based empathies?

"Please Continue": Empathy and the Limits of Thought Experiments

In a now famous series of studies, psychologist Stanley Milgram investigated the influence of authority on moral behavior (Milgram 1963).[17] Milgram wondered whether the directions or behavior of others could strongly influence a person's own moral choices. In the most famous version of Milgram's studies, Milgram posed as an experimenter interested in punishment and learning. His subjects were told that they would be paired with another subject who would play the role of the learner. In reality, the learners were in on the study, not actual subjects, and they were never in any danger.

Milgram's actual subjects were instructed to read a word aloud and wait for the correct response from the "learner," whom they were told was actually the real subject of the study. If the learner answered correctly, they were to move on to the next word pairing. However, if the learner answered incorrectly, the subject was instructed to deliver an electric shock to the learner before moving on. The shocks were, of course, not real. However, Milgram recorded a series of realistic cries, yells, and pleas that were played when subjects delivered these (fake) shocks. These were played over a speaker at specific times. The shocks also grew in severity:

> The instrument panel consists of 30 lever switches set in a horizontal line. Each switch is clearly labeled with a voltage designation that ranges from 15 to 450 volts. There is a 15-volt increment from one switch to the next going from left to right. In addition, the following verbal designations are clearly indicated for groups of four switches going from left to right: *Slight Shock, Moderate Shock, Strong Shock, Very Strong Shock, Intense Shock, Extreme Intensity Shock, Danger: Severe Shock.* (Two switches after this last designation are simply marked *XXX*).[18]
>
> (Milgram 1963, 373)

The real experiment was in seeing how far subjects would be willing to go in terms of delivering increasingly violent shocks to the learners as part of their role in an experiment. If subjects expressed hesitation, Milgram was ready to respond with scripted lines meant to nudge them into compliance, including:

> Prod 1: Please continue. or Please go on.
> Prod 2: The experiment requires that you continue.
> Prod 3: It is absolutely essential that you continue.
> Prod 4: You have no other choice, you must go on.
>
> (1963, 374)

Milgram's prods were effective nudges (i.e., they don't *force* the subject to continue delivering shocks, but they do add social pressures meant to incentivize staying involved).[19] What does all of this have to do with empathy and perspectival thought experiments?

What do you think you should do if you were in Milgram's study? Do you think you would be willing to do as *all* of his subjects did and deliver shocks at the *Intense Shock* level? Would you be one of the 65 percent of subjects who delivered every single shock (including the second *XXX* shock)? Do you think you could do this even though, at 300 volts, "the victim kicks on the wall and no longer provides answers to the teacher's multiple-choice questions" (Milgram 1963, 375)? Most of us think that we wouldn't be capable of killing someone just because a person in authority firmly, but politely, nudges us to. However, if Milgram's data is accurate, then it suggests, quite strongly, that we would be wrong about what we would judge and what we would do in these contexts.[20] This kind of failure is instructive.

On the one hand, it minimally suggests that we can be wildly wrong about the accuracy of our theory-theory empathizing even when the person we're empathizing with is ourselves! However, perspectival thought experiments don't aim for theory-theory empathizing, they aim for in-their-shoes or simulation based forms of perspective taking. In this case, I'm asking you to imagine an experience in a possible future: what would you do if you were in a situation like this? It does appear as if people struggle to accurately imagine what they themselves would actually do (as opposed to what they might hope they would do) in a Milgram-style situation. This kind of task requires that we imaginatively reconstruct some as-yet-unencountered situation. For reasons we'll see shortly, it might be impossible for us to imaginatively construct nearly all interesting moral situations.[21]

To see how deeply these problems go, it's helpful to see that we're not just bad at imaginatively *creating* novel situations but we're also quite bad at imaginatively *recreating* past experiences. One might think, for example, that recreating past experiences might be something we're better at (since we've actually lived through the experience) and might, therefore, be something that could allow researchers to salvage essential pieces of the Thought Experiment Paradigm. There is evidence that "phenomenological experience differs markedly when they are predicting versus remembering emotion" (Levine et al. 2018) and that when predicting our emotional responses "research has shown that actual emotions are typically less intense than expected" (Lench, Safer, & Levine 2011). We might be able to ask subjects to think back to some life or death situation they've faced and add moral dilemmas to these recalled situations to gather better data.

We have some reason to think that even this simpler task is beyond our reach in most cases and for the same reasons as why we fail at predictive simulation. Linda Levine, a psychologist at the University of California–Irvine, has conducted a series of studies that cast our ability to empathize with our past selves into doubt. We might think that our ability to imaginatively reconstruct and re-experience our own past mental states might fade with time. We might not be surprised if we can't accurately re-experience what it was like for us when we were ten or eighteen. However, we might think that our ability to re-experience something important that happened only a few months ago should be stronger.

In one study (1997), Levine studied people's ability to remember how they felt after their preferred presidential candidate withdrew from an election.[22] Levine's results suggest that even after only four months following an emotionally laden experience, our ability to accurately recreate and re-experience these emotional states is fragile:

> Systematic distortions were found in the recall of past emotions as a function of changing appraisals. Supporters who remained loyal to [the candidate] underestimated how sad and angry they had reported feeling after [the candidate] first withdrew from the race and overestimated initial reports of hope. Supporters who planned to vote for other candidates after [the candidate] withdrew, but who later wished [the candidate] had been elected, underestimated how angry they had been when [the candidate] first withdrew.
>
> (174)

Why might this be? In her work on our ability to imagine future emotional responses (something many perspectival thought experiments require their subjects to do), Levine argued that her research suggests that people are bad at this task because "people fail to take into account the surrounding context that mitigates the intensity of their reactions after an event" (Lench, Safer, & Levine 2011).

Levine and her colleagues are not alone in having arrived at this conclusion. Both philosophers and psychologists have begun to wonder whether the sort of questions, and more importantly the sort of *task*, that perspectival thought experiments require subjects to engage in make for poor results. In particular, it seems that our inability to imaginatively place ourselves into or to recreate the contextual features of a situation play a strong role in that explanation:

> Decisions on whom to kill or save are scarcely made by the average person, and participants in psychological experiments only rarely, if ever, have faced such problems in their life. Therefore, when asked about their predicted behavior, people are forced to use their intuitions instead of relying on past experiences ... Such intuitions are, however, not a perfect predictor of what people will really do, as their behavior is often affected by the context in which such choices are made.
>
> (Bialek et al. 2019)

The problem with perspectival thought experiments is connected directly to our trouble imaginatively accounting for contextual features of a situation. When looking *back* at our experiences, we're rarely in the same context, at least when it comes to experimentally interesting situations. The same problem applies to prospective situations. Our inability to imaginatively account for contextual features of a situation thus leads not only to bad predictions (i.e., about Milgram-style situations) but also inaccurate empathizing. To see why this problem is more than a practical problem but also a conceptual one, we need to say more about an issue first identified by a philosopher named Peter Goldie with regard to the nature of perspective taking and what it means to shift into one.

Perspectives have a way that they *feel* to the person experiencing them. They have a phenomenology, a "what it's like" to be in them. Perspectives, whatever else they are, can be said to contain at least two kinds of elements that contribute to their phenomenology: conscious (sometimes called doxastic) and non-conscious (subdoxastic) elements. Doxastic elements of a perspective refer to the explicitly conscious, usually propositional, attitudinal features of an experience.

When I look outside my window I can see downtown San Jose. I see construction on the corner of my block. I see traffic, low mountains on the horizon, and a blue sky. Doxastic elements include all of our sense modalities (this smells like garlic, that feels rough, etc.). All of these are conscious, doxastic, elements of all of our current experiences. Part of what it means to *see things* from my point of view is to token the propositional attitudes I'm tokening. However, there is more to a perspective than just its conscious, propositional, doxastic elements. To speak of doxastic elements is often shorthand for speaking of how a person conceptualizes their sensory stimuli (the concepts that they apply to their experience to help them make sense of it).

Subdoxastic elements of a perspective are equally important as the doxastic, though often they can be harder to notice. Subdoxastic features of an experience *affect* conscious experience but are not themselves a part of our conscious experience. To call something subdoxastic is to say that it is non-conscious. For example, our emotions (positive, negative, and even neutral) can trigger suites of heuristics and biases that affect our cognition in fairly well-understood ways. Each of these emotional heuristics can change how we consciously experience the situations around us (Alloy & Abramson 1979; Kahneman 2003; Ramirez 2017). Feeling happy? Positive affect can make you more likely to rely on stereotyped information about others (Lount 2010) while broadening the scope of attention and thought–action repertoires (Fredrickson & Branigan 2002). Feeling sad? Depressive realism can make your judgments about task duration more accurate than they would be (Moore & Fresco 2012).

To say that these effects are subdoxastic is to say that they are not present as a component of a subject's conscious experience. Positive affect biases can have their effects, insofar as they have them at all, by staying in the background of a subject's conscious experience.[23] In order to be subdoxastic, they *cannot* be a part of the subject's conscious experience. Depressive realism wouldn't work if one consciously thought to oneself, "I'm sad right now, let's make predictions about task duration." Other elements of a situation (like time pressure) or the triggering of implicit biases (Avenanti,

Sirigu, & Aglioti 2010) can alter the way we think and affect our rational/ moral calculus in ways that are not typically a part of conscious decision-making.

Think back to Levine's research on reconstructing past emotional experiences. In those cases, we very quickly get worse at recreating emotionally laden past experiences for subdoxastic reasons. Because our current situations are likely to contain different subdoxastic features than our past situations, we have difficulty accurately recreating their effects when we imaginatively reconstruct them.

Despite all of this, it does sometimes seem as if we are able to engage in perspective taking. Why might this be? One answer might be that, in a limited class of cases, we are in fact able to place ourselves into someone else's (or our own) shoes. Peter Goldie refers to these as "base cases" of empathy:

> Base cases are those cases where "(i) there are no relevant differences in the psychological dispositions of A, the person attempting to empathize, and of B, the target of the attempt; in particular, both A and B are minimally rational; (ii) there are no relevant non-rational influences on Bs psychological make-up or decision-making processes; (iii) there is no significant confusion in Bs psychological make-up; and (iv) B is not faced with a psychological conflict, such as having to make a choice between two or more alternatives where it is not clear to B which alternative is to be preferred.
>
> (Goldie 2011)

Suppose I am to imagine what it will be like for me to go into my office during a typical post-Covid Monday. Further, let's assume that this coming Monday isn't very different from a typical post-Covid Monday. Suppose further that I am not now (as I type this) in the grip of any especially important subdoxastic situational pressures (time pressure, coffee anxiety, etc.), then I ought to be able to successfully engage in putting myself into my own Monday office shoes fairly accurately. Such a case would be a base case for empathy and it could work not only for putting myself into my own future Monday perspective but also possibly into someone else's position as well. Base cases, then, are those where the differences between the empathizer and their target are irrelevant because idiosyncratic features of each agent don't impact the goal of empathy. How might that work?

There are many instances in which two people might stand in a base case relation to one another. I've argued elsewhere (Ramirez 2017) that John Rawls' own classic Original Position thought experiment (Rawls 1971) is an attempt to create a base case for us to make universalizable decisions about how to distribute primary social goods. Part of the function of the veil of ignorance, perhaps its chief function, is to render inert any idiosyncratic subdoxastic elements of experience such that we can all engage in the same process of rational deliberation. This is why Rawls thinks that we would all arrive at the same conclusions. Almost all interesting cases are not base cases, however, and Goldie can help us explain why attempting to imaginatively place ourselves into someone else's shoes (or even into our own past and future shoes) goes awry in all such cases:

> As attempt at perspective-shifting to Bs psychology will have to involve taking on those aspects of Bs characterization that differ from her own whilst at the same time not being conscious of them as such. The reason for this is that the typical role of these dispositions is passive or in the back ground in the sense that our conscious thoughts and feelings that feature in our deliberations are shaped by, but are not directed towards, these dispositions.
>
> (Goldie 2011)

In this case A and B need not be different people. They could represent one and the same person attempting to shift into their own past or future states (when they are not in a base case relation). The essence of the problem is that constructing a situation in our minds is, by definition, a conscious process. While this isn't *necessarily* problematic for doxastic elements of a perspective, this ineliminable feature of the imagination renders it impossible to recreate subdoxastic elements of a perspective:[24]

> A cannot, as part of a consciously willed project, keep Bs characterization in the non-conscious background in her imaginative exercise of wondering what B will decide to do in a certain situation. A will be obliged, in trying to shift to Bs perspective, to treat Bs characterization through the theoretical or empirical stance, as one typically does when considering the role of character in explaining or predicting other peoples' decisions, actions, and so on ... this produces a fundamentally distorted model of Bs thinking.
>
> (Goldie 2011)

Here I think we can see most clearly *why* psychologists who study memory and the imagination continue to find that the imagination's ability to construct and place ourselves into past or future events is limited. When we ask the imagination to construct a perspective to which we do not currently stand in a base case with, we ask it to do the impossible. Similarly, when researchers within the Thought Experiment Paradigm ask subjects to do the same, they set themselves, and their subjects, up for failure.[25] At best, when it seems like subjects say that they'll flip the switch, torture the terrorist, or push the man onto the path of the trolley subjects are not really answering the questions they're being posed.

Instead of giving researchers a response based on successful in-their-shoes or simulation based perspective taking, what they're really doing is trying to predict what they *think* they might do. In other words, what subjects are really doing when they respond to perspectival thought experiments is engaging in either descriptive or normative theory-theorizing. This is bad for two reasons. First, we've already seen that our predictions about our future behavior will often radically fail to track our actual behavior (as they did in Stanley Milgram's studies). Second, it overestimates our ability to imaginatively acquaint ourselves with *what it would be like* to be in those situations.

If, as adherents of the Thought Experiment Paradigm appear to commit themselves to, our responses to perspectival thought experiments really do tell us what we would do (or what we would believe is justified for us to do), then it makes sense that such experiments can be assigned a probative role. But they *can't* do this. As a result, whenever perspectival thought experiments are used probatively, an important mistake is being made and progress is being hampered. That being said, all is not lost.

Many philosophers and psychologists have begun to argue that "recent advances in immersive virtual environment technology allow for ... studies to be conducted in artificial, yet realistic, 3-D digital worlds" (Navarrete et al. 2012, 365). Recall that the problem with perspectival thought experiments is ultimately grounded in the impossibility of imaginatively recreating the subdoxastic elements of experience. Because imagining what it might be like to be confronted with a trolley problem is necessarily a conscious project, it is impossible to accurately imagine how subdoxastic aspects of that situation would affect us. We can't consciously add in the unconscious influences on our beliefs or perceptions without changing their nature. To do so implies a shift from in-their-shoes empathizing over to theory-theorizing forms of empathy. However, when we build a virtual or augmented reality world and then expose people to events in that world, they can react to the subdoxastic elements of that world in a much more realistic way.

Over the next several chapters, I'll develop a more rigorous theory about the elements of a virtual experience, but there is reason to be optimistic about using virtual reality analogues of perspectival thought experiments to make philosophical and psychological progress:

> Such "virtual reality" (VR) experiments reveal a high degree of congruence with behavior observed in typical laboratory settings, but with more experimental control ... . For example, Slater and colleagues (2006) found that participants in a VR simulation of Milgram's classic obedience experiments performed almost exactly the way one would expect them to on the basis of the results of the original studies, delivering dangerous shocks to a virtual person following instructions from a virtual experimenter, and with autonomic responses that measured emotional arousal covarying with both the severity of the shock delivered and with proximity to the virtual subject. Likewise, Dotsch and Wigboldus (2008) found that implicit attitudes and autonomic arousal predicted participants' physical distance from an ethnic minority target

in a virtual world, consistent with previous work linking nonconscious attitudes and anxiety to behavioral discrimination.

(Navarrete et al. 2012)

We'll have reason to question some of the claims that psychologists make about virtual reality (not all VR simulations of moral dilemmas are well-designed, not all well-designed VR simulations are ethical to use). Though I'll be more sanguine about what VR simulations can ultimately achieve (hint: simulation empathy is impossible even in VR), this does not mean that they don't represent a significant improvement over the Thought Experiment Paradigm.

Notes

1 Henry Shue posed this now famous thought experiment in his (1978) article called "Torture."
2 William Swann and his colleagues (2010), for example, investigated whether Spaniards would be more or less willing to sacrifice their own lives to save other Spaniards or non-Spaniards in trolley problem situations. They found that in-group/out-group effects like this can matter to their subjects (i.e., being more willing to save fellow Spaniards or even fellow Europeans more than hypothetical Americans). So it might very well matter who we picture when we imagine these situations.
3 Thought experiments, even moral thought experiments, long predate the 1970s! David Hume, for example, famously used thought experiments in connection with arguments for an "ideal observer" version of moral sentimentalism (Hume 2007). I reserve the phrase "Thought Experiment Paradigm" to describe the explosion of largely empirical moral psychological research utilizing moral vignettes as their primary stimulus/variable. Psychologists, philosophers, and neuroscientists have routinely appealed to moral vignettes of this kind to draw conclusions within what I'm calling the Thought Experiment Paradigm.
4 What do I mean by "complete successfully" in this context? As we'll see, I mean this quite literally. We can't complete the trolley problem if we imagine the trolley problem as requiring us to do what it seems to require us to do: to imagine ourselves in a situation where we must either pull a switch and kill one or not pull a switch and allow five to die. While it certainly might seem as if we can imagine all of this, I'll argue that we're fundamentally mistaken about this.
5 Importantly, it matters whether a thought experiment asks us the question "what would you do?" as opposed to something like "what should someone do?" The first question is distinctly perspectival, the second is not. I'll say more about this distinction later in this chapter. For now, it matters that only the perspectival question would generate new information (new reasons) and thus only the perspectival question can serve a probative function. It's not surprising, then, that philosophers and psychologists tend to interpret the trolley problem as a perspectival thought experiment. To ask the more straightforwardly normative question would be answerable without really imagining anything at all (there isn't any experiment in such a "thought experiment"). To predict our future behavior is not to test it or run an experiment at all, it is to apply a background theory to a situation. Given the way that philosophers and psychologists have put perspectival thought experiments to use, this cannot be what they aim to do.

6 Individuals with a condition known as aphantasia form a clear limit case for perspectival thought experiments (Keogh & Pearson 2018). Such individuals are incapable of, or have significant difficulty with generating mental imagery of any kind. Individuals with aphantasia thus ought not be able to complete perspectival thought experiments, though they ought to be capable of the much less visually demanding normative question (what should you do?). I do not know of any empirical work that assesses this issue but the results of such an experiment would be illuminating.

7 In fairness to Liao et al., they argue that the results of their (2012) research ought to undermine (but not eliminate) our confidence in the usefulness of moral intuitions. I tend to agree with them on this, though as we shall see, for different reasons. My issue with the kinds of moral intuitions that they elicited in their study isn't that they're folk-moral intuitions (e.g., that they're the moral intuitions of those untrained in philosophical ethics), it's that these intuitions were generated using perspectival thought experiments. By the end of this chapter, it should be clear that much better data can be generated by offloading the perspectival task from the subject. We get much better data on moral intuitions (genuinely useful empirically and philosophically) by placing subjects into a VR or AR simulation of the trolley problem than by asking subjects to imagine it.

8 As this example shows, a single thought experiment can play more than one role. In this case, rhetorical power might be the dominant role someone might ask this perspectival thought experiment to play. However, it can certainly be intended to play a probative or elucidatory role. I'll argue in a later chapter that virtual reality versions of these thought experiments, which aim to play both rhetorical and probative roles, will often function as unethical nudging devices.

9 We seem capable of empathizing via mirroring not only with other people but also with non-human animals, fictional persons, and even inanimate objects like animated toasters.

10 Interestingly, and importantly, even though mirroring empathy is non-cognitive, it has been shown to respond to internalized biases. For example, we more strongly mirror the pains of those whom we identify as in our racial in-group than with those we identify as in our out-group (Avenanti, Sirigu, & Aglioti 2010).

11 As we'll see, while "mind-reading" (what I'll call "theory-theory") empathy is dimensional, its cognitively effortful nature makes it distinct from mirroring empathy. In other words, while some people are much better than others at figuring out how others think and feel, engaging in this form of empathizing requires conscious effort. Mirroring empathy does not. It is, as I noted, a non-cognitive process.

12 I say "controversially" to note that those who are part of the neurodiversity movement argue that spectrum capacities like mirroring empathy (and other forms of empathizing) are proof of differences in capacity but not by themselves proof of pathology. For example, neurodiversity proponents will argue that people on the autism spectrum are not disordered (i.e., that autism is not a disease) but just another way in which people can be born (Ramirez 2016).

13 The term psychopathy is also incredibly controversial. I (Ramirez 2015) intend to pick out the condition characterized by a high score on a diagnostic tool known as the Psychopathy Checklist (the PCL-R) and not sociopathy as defined in the Diagnostic and Statistical Manual of Mental Disorders (APA 2013).

14 Hume might be asking us to instead think about how we have felt in the past under similar circumstances and to correct our present emotional impressions by adjusting for these past experiences. If that's how we interpret Hume then this points to yet another form of empathy: theory-theory empathy.

15 Keeping in mind what I noted in note 12, I use the phrase "autism spectrum disorder" without endorsing the view that high functioning persons with autism are disordered at all.

16 That is, of course, unless mirroring empathy plays an important role in the context of the thought experiment itself!

17 Situationists about moral psychology believe that we can learn a lot about the nature of character, especially moral character from these studies (Doris 1998; Doris 2002; Doris 2006). Although questions about the nature of moral character and its relation to empirical research like Milgram's are slightly outside of the scope of this chapter, I think that we can learn just as much about virtual and augmented reality moral psychology by looking at replications of Milgram's research using these technologies. I say more about Milgram and these replication attempts in the next chapter.

18 Emphasis is original.

19 In Chapter 5, I lay out the conditions under which VR nudges can be unethical to develop or deploy.

20 Although exact replications of Milgram's experiments are now off-limits for ethical reasons, there have been several replications of Milgram's experimental paradigm that suggest his results weren't unique to his subjects or the norms of people in the US in the 1960s (Burger 2009; Dolinski et al. 2017; Slater et al. 2006).

21 I say "interesting" here because it will turn out that we are quite capable of in-their-shoes empathizing about cases that are almost exactly like the situation we currently find ourselves in. For example, I'm currently sitting in San Jose, at my kitchen table, writing on a laptop. It will be possible for me to successfully complete a perspectival thought experiment that asks me to imagine that I'm sitting at a desk in an apartment in San Jose that is very much like my own and which is not subject to conditions that I'm not also subjected to. However, when we think about the kinds of situations that are typically used by those within the Thought Experiment Paradigm, it's highly unlikely that any subject will be situated in such a way that they are actually capable of completing it. More on all of this in a moment.

22 Writing this in 2020–2021, this research feels especially important!

23 In other words, the subject does not need to realize that she is feeling good and therefore that she is now more likely to rely on stereotypes when she makes judgments.

24 I say that it isn't necessarily problematic because it often will turn out to be! For example, when I look outside my window and see downtown San Jose, the doxastic character of my experience isn't simply the tokening of the belief "that over there is downtown San Jose." Suffused in that tokening are a wide range of historical, emotional, subdoxastic features. What I see when I look at downtown San Jose is likely colored by these features and expectations. That coloring can matter when we recreate perspectives imaginatively (Noe 2004).

25 One of the central issues in Chapter 5 will be to provide design guidelines for how to design virtual reality analogues of perspectival thought experiments so that they are as ecologically valid as possible while avoiding unnecessary subject trauma.

References

Allman, J., & Woodward, J. 2008. What are moral intuitions and why should we care about them? A neurobiological perspective. *Philosophical Issues*, 18 (1), 164–185.

Alloy, L.B., & Abramson, L.Y. 1979. Judgment of contingency in depressed and nondepressed students: Sadder but wiser? *Journal of Experimental Psychology*, 108, 441–485.

APA (American Psychiatric Association). 2013. *Diagnostic and statistical manual of mental disorders, 5th edition.* Washington, DC: APA.

Andrews, K. 2008. It's in your nature: A pluralistic folk psychology. *Synthese*, 165 (1), 13–29.

Avenanti, A., Sirigu, A., & Aglioti, S.M. 2010. Racial bias reduces empathic sensorimotor resonance with other-race pain. *Current Biology*, 20 (11), 1018–1022.

Battaly, H.D. 2011. Is empathy a virtue? In Coplan, A. & Goldie, P. (Eds.), *Empathy: Philosophical and Psychological Perspectives.* Oxford: Oxford University Press, 277–301.

Białek, M., Turpin, M.H., & Fugelsang, J.A. 2019. What is the right question for moral psychology to answer? Commentary on Bostyn, Sevenhant, and Roets (2018). *Psychological Science*, doi:10.1177/0956797618815171.

Bloom, P. 2014, September 10. Against empathy. *Boston Review: A Political and Literary Forum.* Retrieved from http://bostonreview.net/forum/paul-bloom-aga inst-empathy

Brown, J. 2011. *The laboratory of the mind.* 2nd ed. London: Routledge.

Bruers, S., & Braeckman, J. 2014. A review and systematization of the trolley problem. *Philosophia*, 42, 251–269.

Buckwalter, W., & Turri, J. 2016. In the thick of moral motivation. *Review of Philosophy and Psychology*, 8, 433–453.

Bunzl, M. 1996. The logic of thought experiments. *Synthese*, 106 (2), 227–240.

Burger, J. 2009. Replicating Milgram: Would people still obey today? *American Psychologist*, 64, 1–11.

Choi, C.Q. 2007, June 17. Study: People literally feel pain of others. *Livescience.* Retrieved from https://www.livescience.com/1628-study-people-literally-feel-pain. html

Colle, L., Baron-Cohen, S., & Hill, J. 2007. Do children with autism have a theory of mind? A non-verbal test of autism vs. specific language impairment. *Journal of Autism and Developmental Disorders*, 37 (4), 716–723.

Davis, M. 2012. Imaginary cases in ethics: A critique. *International Journal of Applied Philosophy*, 26 (1), 1–17.

Di Nucci, E. 2012. Self-sacrifice and the trolley problem. *Philosophical Psychology*, 26 (5), 662–672.

Dolan, M., & Fullam, R. 2004. Theory of mind and mentalizing ability in antisocial personality disorders with and without psychopathy. *Psychological Medicine*, 34, 1093–1102.

Dolinski, D., Grzyb, T., Folwarczny, M., Grzybała, P., Krzyszycha, K., Martynowska, K., & Trojanowski, J. 2017. Would you deliver an electric shock in 2015? Obedience in the experimental paradigm developed by Stanley Milgram in the 50 years following the original studies. *Social Psychological and Personality Science*, 8 (8), 927–933.

Doris, J.M. 1998. Persons, situations, and virtue ethics. *Noûs*, 32, 504–530.

Doris, J.M. 2002. *Lack of character: Personality and moral behavior.* Cambridge: Cambridge University Press.

Doris, J.M. 2006. Out of character: On the psychology of excuses in the criminal law. In La Follette, H. (Ed.), *Ethics in practice.* Oxford: Blackwell.

Einstein, A. 1920. *Relativity: The special and the general theory*. Lawson, R.W. (translator). London: Methuen. Retrieved from https://archive.org/stream/relativi tythespe00einsuoft#page/n3/mode/2up

Foot, P. 1978. The problem of abortion and the doctrine of double effect. In *Virtues and vices and other essays in moral philosophy*. Oxford: Clarendon Press.

Fredrickson, B.L., & Branigan, C. 2002. Positive emotions broaden the scope of attention and thought-action repertoires. *Cognition and Emotion*, 19 (3), 313–332.

Gendler, T. 2007. Philosophical thought experiments, intuitions, and cognitive equilibrium. In French, P.A. & Wettstein, H.K. (Eds.), *Philosophy and the empirical*, Oxford: Blackwell, 68–89.

Goldie, P. 2011. Anti-empathy. In Coplan, A. & Goldie, P. (Eds.), *Empathy: Philosophical and psychological perspectives*. Oxford: Oxford University Press, 318–330.

Goldman, A.I. 2011. *Two routes to empathy*. In Coplan, A. & Goldie, P. (Eds.), *Empathy: Philosophical and psychological perspectives*. Oxford: Oxford University Press: 31–44.

Goldman, A.I., & Jordan, L. 2013. Mindreading by simulation: The roles of imagination and mirroring. In Baron-Cohen, S., Lombardo, M. & Tager-Flusberg, H. (Eds.), *Understanding other minds: Perspectives from developmental social neuroscience*, 3rd ed. Oxford: Oxford University Press, 448–466.

Greene, J.D., Sommerville, R.B., Nystrom, L.E., Darley, J.M. & Cohen, J.D. 2001. An fMRI investigation of emotional engagement in moral judgment. *Science*, 293 (5537), 2105–2108.

Greenspan, P. 2003. Responsible psychopaths. *Philosophical Psychology*, 16 (3), 417–429.

Haidt, J., & Bjorklund, F. 2007. Social intuitionists answer six questions about morality. In Sinnott-Armstrong, W. (Ed.), *Moral psychology, vol. 2: The cognitive science of morality*. Cambridge MA: MIT Press, 181–217.

Hobson, P. 2002. *The cradle of thought*. Oxford: Oxford University Press.

Huebner, B., & Hauser, M. 2011. Moral judgments about altruistic self-sacrifice: When philosophical and folk intuitions clash. *Philosophical Psychology*, 24 (1), 73–94.

Hume, D. 1738/2007. *A treatise of human nature*. Norton, D.F. & Norton, M.J. (Eds.). Oxford: Oxford University Press.

Iacoboni, M. 2009. Imitation, empathy, and mirror neurons. *Annual Review of Psychology*, 60, 653–670.

Kahneman, D. 2003. Maps of bounded rationality: Psychology for behavioral economics. *American Economic Review*, 93 (5), 1449–1475.

Kant, I. (1785/1998). *Groundwork of the metaphysics of morals*. Gregor, M. (Ed.). New York: Cambridge University Press.

Keogh, R., & Pearson, J. 2018. The blind mind: No sensory visual imagery in aphantasia. *Cortex*, 105, 53–60.

Lench, H.C., Safer, M.A., & Levine, L.J. 2011. Focalism and the underestimation of future emotion: When it's worse than imagined. *Emotion*, 11, 278–285.

Levine, L.J. 1997. Reconstructing memory for emotions. *Journal of Experimental Psychology: General*, 126, 165–177.

Levine, L.J., Lench, H.C., Karnaze, M.L., & Carlson, S.J. 2018. Bias in predicted and remembered emotion. *Current Opinion in Behavioral Sciences*, 19, 73–77.

Liao, M.S., Wiegmann, A., Joshua, A., & Vong, G. 2012. Putting the trolley in order: Experimental philosophy and the loop case. *Philosophical Psychology*, 25 (5), 661–671.

Lount, R.B., Jr. 2010. The impact of positive mood on trust in interpersonal and intergroup interactions. *Journal of Personality and Social Psychology*, 98 (3), 420–433.

Machery, E. 2011. Thought experiments and philosophical knowledge. *Metaphilosophy*, 42 (3), 191–214.

Marsh, A.A., Finger, E.E., Fowler, K.A., Cadalio, C.J., Jurkowitz, I.T.N., Schechter, J.C., Pine, D.S., Decety, J. & Blair, R.J.R. 2013. Empathic responsiveness in amygdala and anterior cingulate cortex in youths with psychopathic traits. *Journal of Child Psychology and Psychiatry*, 54 (8), 900–910.

McDonald, N.M., & Messinger, D.S. 2010. The development of empathy: How, when, and why. In Acerbi, A., Lombo, J.A., & Sanguineti, J.J. (Eds.), *Free will, emotions, and moral actions: Philosophy and neuroscience in dialogue*. Vatican City: LIF-Press.

Meynell, L. 2014. Imagination and insight: A new account of the content of thought experiments. *Synthese*, 191, 4149–4168.

Milgram, S. 1963. Behavioral study of obedience. *Journal of Abnormal and Social Psychology*, 67, 371–378.

Moore, M.T., & Fresco, D.M. 2012. Depressive realism: A meta-analytic review. *Clinical Psychology Review*, 32 (6), 496–509.

Myers-Schulz, B., & Schwitzgebel, E. 2013. Knowing that P without believing that P. *Noûs*, 47 (2), 371–384.

Navarrete, C.D., McDonald, M.M., Mott, M.L., & Asher, B. 2012. Virtual morality: Emotion and action in a simulated three dimensional trolley problem. *Emotion*, 12 (2), 364–370.

Nichols, S. 2007. *Sentimental rules*. Oxford: Oxford University Press.

Noe, A. 2004. *Action in experience*. Cambridge MA: MIT Press.

Norton, J. 2004. Why thought experiments do not transcend empiricism. In Hitchcock, C. (Ed.), *Contemporary debates in the philosophy of science*. Oxford: Blackwell, 44–46.

Pastotter, B., Gleixner, S.Neuhauser, T., & Karl-Heinz, T.B. 2013. To push or not to push? Affective influences on moral judgment depend on decision frame. *Cognition*, 126 (3), 373–377.

Ramirez, E. 2015. Receptivity, reactivity, and the successful psychopath. *Philosophical Explorations*, 18 (3), 330–343.

Ramirez, E. 2016. Neurosurgery for psychopaths? The problems of empathy and neurodiversity. *American Journal of Bioethics: Neuroscience*, 7 (3), 166–168.

Ramirez, E. 2017. Empathy and the limits of thought experiments. *Metaphilosophy*, 48 (4), 504–526.

Ramirez, E. 2019. Psychopathy, autism, and basic moral emotions: Evidence for sentimentalist constructivism. In Bluhm, R. and Tekin, Ş. (Eds.), *Bloomsbury companion to the philosophy of psychiatry*. New York and London: Bloomsbury.

Rawls, J. 1971. *A theory of justice*. Cambridge MA: Harvard University Press.

Richell, R.A., Mitchell, D.G.V., Newman, C., Leonard, A., Baron-Cohen, S., & Blair, R.J.R. 2003. Theory of mind and psychopathy: Can psychopathic individuals read the "language of the eyes"? *Neuropsychologia*, 41, 523–526.

Shue, H. 1978. Torture. *Philosophy and Public Affairs*, 7, 124–143.

Simmons, A. 2014. In defense of the moral significance of empathy. *Ethical Theory and Moral Practice*, 17, 97–111.

Slater, M., Antley, A., Davison, D., Swapp, D., Guger, C., Barker, C., Pistrang, N., & Sanchez-Vives, M.V. 2006. A virtual reprise of the Stanley Milgram obedience experiments. *PLoS ONE*, 1, e39.

Swann, W.B., Gómez, A., Dovidio, J.F., Hart, S., & Jetten, J. 2010. Dying and killing for one's group: Identity fusion moderates responses to intergroup versions of the trolley problem. *Psychological Science*, 21 (8), 1176–1183.

Thomson, J. 1976. Killing, letting die, and the trolley problem. *The Monist*, 59, 204–217.

Thomson, J. 1985. The trolley problem. *Yale Law Journal*, 94 (6), 1395–1415.

Ward, J., & Banissy, M.J. 2015. Explaining mirror-touch synesthesia. *Cognitive Neuroscience*, 6 (2–3), 118–133.

3   When Being There Is Not Enough

In the last chapter, I discussed the psychological concept of presence. There, I defined it as the feeling of being physically located in a simulated environment instead of wherever it happens to be that you're located in reality. This chapter focuses on presence and why philosophers, psychologists, and game designers need to go beyond the concept of presence to understand the ethical issues virtual and augmented reality technologies introduce into their work.

When it comes to generating interesting questions about the ethics of building, and using, virtual worlds, it won't be enough for a simulation to generate the feeling of presence in its users.[1] At the very least, we'll need to be more precise about what we mean by the concept of presence. There are different definitions on offer and we'll have to examine which sense of VR and AR presence makes the most sense to use, and it may turn out that philosophers and psychologists are often speaking past each other. As I've noted throughout, almost all of the research conducted on the concept of presence is carried out in VR though, as we'll see, the line between VR and AR is blurry. In some ways, AR presence is easier to induce than VR presence though I'll need to say more about what this difference amounts to. We'll also need to be clearer about how different aspects of a simulation, and its users, can come together to give shape to the experiences users have in them. Take the following cases as examples:

> Samaya is using a virtual reality drafting program as part of her work for an architecture firm. Her current project involves designing a bridge that will span a pair of train tracks. This program includes many virtual tools, which appear to Samaya as floating icons within her field of view, that can be used to modify her designs and test her structures. As she works on a particularly tricky part of the bridge, Samaya adjusts her virtual perspective so that she is floating underneath the bridge itself in order to get to an especially tricky design element. As she's doing this, she automatically reaches out with her hand, instead of her controller, to grab a virtual tool and bumps her hand against her real-life computer monitor.

DOI: 10.4324/9781003042228-3

Giordan is playing a brand new Sci-fi VR shooter called *Firepower*, and he is loving the experience so far. As an experienced gamer, Giordan can deftly maneuver around most of the obstacles the game puts in his way. He's adept at using virtual cover to his advantage to slay his way through the game's levels. Giordan's friends, watching him from his room, and outside the game, often giggle as Giordan tends to physically duck, stand, and lean to avoid taking damage from his virtual enemies. During a particularly intense firefight between Giordan and a techno-zombie horde, he even lets out a victorious scream as he clears the room with a pulse grenade and nearly falls over as he ducks away from flying pieces of techno-zombie gore.

Sam is trying out a new virtual reality exposure therapy tool called Safe-Space. SafeSpace is designed by clinical psychologists to help its users better appreciate, and learn techniques to mitigate, their post-traumatic stress triggers by using a kind of cognitive behavioral therapy delivered via virtual environments. As Sam enters the virtual world and begins to experience the SafeSpace simulation, which has been tailored to Sam's particular triggers, Sam's heart begins to race and palms begin to sweat. At one point, Sam even instinctually leans away from an especially troubling virtual trigger in the simulation and physically moves to avoid contact.[2]

Luis decided to finally give in to peer pressure and buy himself a pair of AR contact lenses. The first thing Luis does is to buy himself a copy of the latest AR gaming craze: Phantasma. Phantasma is a horror-themed AR game that tracks a user's gaze in order to introduce (simulated) spooky audio and visual content into their peripheral vision (the exact nature can be tailored by the user). Luis loves ghost stories and so he chooses the spirit-themed horror experience. As he makes his way through his daily life, the cloud-based software manages to give Luis some powerful scares (strange knocks, blurred movement just outside his line of sight, etc.). Luis very quickly forgets that he's wearing AR contact lenses at all and starts to genuinely believe that his apartment is haunted. Even after removing the lenses, he can't quite shake the feeling that something in his home is "off."

Although Samaya, Giordan, Sam, and Luis all undergo experiences high in what many psychologists would call "presence," only Sam and Luis, I'll argue, have an experience that goes beyond presence and borders on what I'll call throughout this book "virtually real experiences." Virtually real experiences factor prominently not only in this chapter but throughout this book. The fact that, under certain conditions, VR and AR environments can trigger virtually real experiences is an important element in my account of the ethical issues that VR and AR technologies can introduce. But what exactly *is* a virtually real experience and how is it different from other experiences of presence? Why is it that some simulations seem to generate them in us while other simulations can leave us feeling detached even if they seem to have similar content?

For now, let's define virtually real experiences as those experiences that a subject treats, in the moment, *as if* the experience were real. Treating an experience as if it were real is, in itself, a philosophically loaded thing to say about it. As we'll see, virtually real experiences are not "all or nothing" kinds of things. Some experiences can be more virtually real than others while being less virtually real than still others. In this chapter, I aim to set out the conditions under which a virtual experience comes to be experienced as virtually real by its subject. One of my chief tasks will be to distinguish what it is about simulations that generate experiences of presence, in all its senses, from the features that generate virtually real experiences.

I'll argue that the best evidence available to date suggests that the likelihood of a simulation producing a virtually real experience increases or decreases depending on the interactions between three different elements of a simulated experience. Two of these elements, which I'll refer to as context-realism and perspectival fidelity, can be directly tied to design features of a simulation and are under the control of a simulation's creators.[3] The final element focuses not on the simulation or its features but instead on the psychological dispositions of individual users, especially their tendency to experience dissociation and their relative possession of several of the "Big Five" personality traits. In most cases, these subject-side elements aren't under the control of VR/AR developers.[4]

When the right user (i.e., one with the right mix of dissociative tendencies and personality traits) and the right simulation (a simulation high in context-realism and perspectival fidelity) come together, virtually real experiences can result. Virtually real experiences lie at the core of VR's unique ethical potential and its unique ethical risks. In the following chapter, I'll begin exploring the ethical issues generated by VR and AR simulations, games, and experiments especially as they pertain to risks and benefits of causing users to have virtually real experiences. However, first I need to say more about what presence is and why highly present experiences are not always also high in context-realism or perspectival fidelity, and why such present experiences are not necessarily also always virtually real experiences.

Presence and the Elements of a Virtual Experience

In one sense, presence might refer to a very basic proprioceptive psychological experience. It's in our interest to have evolved psychological mechanisms that keep track of where we are and what our options are, what is ready-to-hand in the spaces we find ourselves in, and our location relative to all of these opportunities (Heidegger 1927; Horrigan-Kelly, Millar, & Dowling 2016). This very basic psychological capacity is an aspect of many (arguably all), our conscious experiences and it's working right now, as you read this sentence, letting you know that you are wherever you think you happen to be (Weibel, Wissmath, & Mast 2010). Like all sense modalities, this one is liable to make mistakes and can be subject to illusions of all sorts (Hegedüs et al. 2014; Guterstam, Gentile, & Henrik Ehrsson 2013). We've all awoken from that very vivid dream (or nightmare!) only to realize that we're not where we expected to be; we've all bumped into that unexpected object when we wake up in a foreign bed, mistaking the alien space for a more familiar locale. Usually though, we're pretty good at generating beliefs about where we are, what we're doing, and what parts belong to us. These proprioceptive mechanisms are responsible for that.

Presence is complicated, however, and several theories of what presence is, and what leads us to have a "present" experience, have been proposed. Once we better understand the concept, we can begin to co-opt the psychological mechanisms responsible for presence to more systematically and consistently fool them (i.e., to make simulations capable of ever more presence). Virtual and augmented reality simulations can be designed so that we, much as we do when we have realistic-seeming dreams, come to believe that we're somewhere we're not. However, psychologists themselves disagree about the concept of presence and it can sometimes seem as if they're talking past one another when they write about the psychological elements of presence. Consider the following two conceptions of presence. First, presence is sometimes understood by psychologists in terms of the way a subject experiences a simulation. On that understanding,

> presence in a [virtual environment] is inherently a function of the user's psychology, representing the extent to which an individual experiences the virtual setting as the one in which they are consciously present. On the other hand, immersion can be regarded as a quality of the system's technology, an objective measure of the extent to which the system presents a vivid virtual environment while shutting out physical reality.
> (Cummings & Bailenson, 2016)

As James Cummings and Jeremy Bailenson understand it, immersion is a way of referring to design features and hardware features that help render a simulation more, or less, likely to make a user feel like they're located, or "present," in a virtual space instead of wherever they happen to actually be. The distinctions between hardware features, software features, and user psychology are important and I'll return to them later. However, even in this context, Cummings and Bailenson refer to presence as a kind of successful location illusion. Does a simulation fool a subject into thinking that they're located in (that they're present in) a virtual space? If so, then the simulation has produced a sense of presence and it's likely done this in light of being highly immersive. But this is not the only way of thinking about what it means to be present within a simulation.

Contrast Cummings and Bailenson's understanding of presence with the following conception proposed by psychologists Maria Sanchez-Vives and Mel Slater:

So what is presence? The common view is that presence is the sense of being in a [virtual environment] rather than the place in which the participant's body is actually located … A fundamentally different view is that presence is "… tantamount to successfully supported action in the environment." The argument is that reality is formed through actions, rather than through mental filters and that "… the reality of experience is defined relative to functionality, rather than to appearances". The key to this approach is that the sense of "being there" in a [virtual environment] is grounded on the ability to "do" there.

(Sanchez-Vives & Slater 2005)

Sanchez-Vives and Slater introduce another sense of presence, we might call it "doing there" presence, which stands in contrast to the more traditional "being there" sense of presence. Doing there presence, on their view, is a complex property. It seems to both ground the kind of presence Cummings and Bailenson mention, but is itself caused by the features of a virtual environment that generate the feeling of being able to manipulate virtual spaces in the same ways that we're used to manipulating real spaces. Location illusion enabled by action illusion. Successfully fooling someone into feeling located in a space, they argue, depends on making them believe that they can manipulate those spaces in the ways that they have come to learn to do more generally, in the real world.

There's much to say here. For the moment, it's enough that we distinguish between those features of a simulation that create the feeling of being *inside* a virtual world from those features of the simulation (and the hardware that runs it) which provide a successful *illusion of acting* in that environment. Both sets of features are related, I'll argue, and we can make better sense of these elements by distinguishing between elements of a simulation's *content* (what I'll refer to as a simulation's context-realism) from the structural features of a simulation that create a user's perspective within the simulated world (what I'll call perspectival fidelity). "Doing there" and "being there" senses of presence depend on both elements in different ways.

Still others have taken a fundamentally different approach to theorizing presence. They, for example, define presence in terms of different psychological capacities, focusing less on a location or action illusion and more on theory of mind. For example, some understand presence as "the process of perspective taking, the mental simulation of a situation by placing oneself in the shoes of another via imagination" (Ahn et al. 2016). This way of thinking about presence is linked to what we said in the last chapter about empathy. Presence, on Ahn et al.'s understanding, involves a simulation's ability to create, for one person, the experience of what it is like to be in someone else's shoes (what we called "in-their-shoes" empathy).

It isn't clear from Ahn et al.'s definition whether *accuracy* is required in this conception of presence (that is, whether our feeling of what it would be like to be in someone else's shoes should accurately track what it would actually feel like to us to be in someone else's position). As we'll see in Chapter 5, ethical issues arise once we use VR and AR simulations for this kind of "empathy enhancement." Here too we can ask about the hardware and software elements that might make some simulations more likely than others to generate perspective-taking illusions in the sense that Ahn et al. have in mind.

For now it's enough to parse out what seem to be a few important elements that these senses of psychological presence have in common:

1   Successful location illusion (for "being there" presence)
2   Naturalistic action affordability (for "doing there" presence)
3   Perspective-shifting (for empathizing presence)
4   The hardware and software features of a simulated world that allow for 1–3[5]

Presence, we might say, is a way of speaking about a property that some simulations have for inducing within a user the (illusory) belief that they're located in a simulated world and are now occupying a perspective other than their own. If presence means something along these lines then Samaya, Giordan, Sam, and Luis were all present during their simulated experiences. But what kind of presence were they experiencing? In virtue of Samaya's reaching out in the real world for a tool that only exists in her simulation, we can infer that she believed (if only temporarily) that she was located within the virtual space of her engineering simulation. Following Sanchez-Vives and Slater, her naturalistic reaching for virtual tools suggests that she was experiencing both "doing there" and "being there" senses of presence.

In light of the fact that Giordan was bobbing and weaving in the real world to avoid attacks by simulated techno-zombies, we can infer that he too believed (if only temporarily) that he was inside the world of his game. He was "being there" in the virtual game space. On the assumption that the VR headset also accurately tracked head movement so that he was successfully avoiding enemies by moving around, he may very well have also experienced "doing there" presence to some degree (made even stronger if he was using a gun-shaped peripheral instead of a joystick or keyboard – more on this in a moment).

Given Sam's highly visceral reactions to the simulated triggers, we can also say that Sam likely believed that those triggers were present in both the "being there" and "doing there" senses. That Sam moved away instinctually from triggers and that Sam's body reacted viscerally in ways that largely mirror how bodies typically respond to real-life triggers suggests that something more was going on than just the sense of being in a virtual space. Although all three simulations seemed to successfully create the feeling of presence in their users, I believe that Sam's experiences show us that simulations can involve effects that go beyond the experiences of "being there" and "doing there" senses of presence. Similarly, Luis' responses to simulated hauntings not only elicited "pretend" fear but seemed to elicit "as if it were real" fear. We can ask further questions based on the fact that long after the simulation itself had ended, Luis appeared to act as if the simulated hauntings gave him real information about his apartment and its possible (new) inhabitants.

Sam and Luis don't seem to believe merely that the triggers in their simulation were present, that is, Sam and Luis seem to think not only that they and their triggers were located in a virtual space but instead that they were really there (even after consciously ending the simulation in Luis' case). Sam's responses to those virtual triggers are the responses of someone acting as if the triggers were real. Luis' continued belief that something in his apartment is "off" also suggests that his simulated experiences have affected his real-life thoughts, beliefs, and actions.

Contrast these responses to the way that Giordan acted while playing his VR game. While Giordan was utilizing the movement afforded to him in virtual space to avoid being killed by techno-zombies, Giordan doesn't appear, even for a moment, to act as if he's really in danger or as if he's been destroying real people or monsters for over an hour. Samaya's simulation allowed her to hang in mid-air underneath the bridge she was designing. While the simulation clearly was able to make her feel like she was located in a virtual space, Samaya doesn't treat that virtual space as if it were real, as if she were really floating beneath a bridge and in danger of falling. Unless Samaya is especially fearless, we would expect her to exhibit more fear, stress, and physiological arousal if she actually believed she was hanging underneath an unfinished bridge.

We ignore the distinctions between these experiences at our peril. Highly immersive experiences capable of generating presence are philosophically and psychologically interesting. For game designers, knowing how to make a simulation more immersive, and thus more likely to generate presence, is part and parcel of creating a successful game. However, these simulations are capable of generating experiences that include not only presence but which can also seem virtually real to their users, and doing *this* could cause ethical and commercial problems. Simulations that do this (that create virtually real experiences) introduce new ethical issues that have been largely ignored in both the psychological and philosophical literature on virtual experiences.

A virtually real experience, like a highly present experience, involves a kind of successful illusion. Where a present experience requires successfully having a user believe they're located inside a virtual space (even if only temporarily), virtually real experiences require successfully having users believe that virtual events and virtual objects are real events or real objects and for their bodies to respond appropriately (even if only temporarily). Virtually real experiences, we can say, are those experiences which, while in the grip of the simulation, are responded to (physiologically, psychologically, behaviorally) as if they were real experiences.

A simulation's ability to produce virtually real experiences in users is not an "all or nothing" thing. Different driving simulations may get our heart racing a little or a lot depending on the user and the simulation's structure. This gives us some reason to think that we're treating these experiences as virtually real to varying degrees. It also includes experiences that go beyond "being there" and "doing there" senses of presence. Virtually real experiences, according to the best available data at the moment, turn out to depend on three different factors that take us beyond location and action illusions. My focus, in this chapter, is to map out the counters of these elements and to explain how this evidence ought to inform our thoughts about VR and AR design.

These elements, which I now turn to, include not only aspects of a simulated world's narrative but also include features of that world's design space and of each user's individual psychological traits. I'll begin by discussing the design elements (both in terms of hardware and software) that can affect the likelihood of a simulation producing a virtually real experience in its users. Once we understand those elements, I'll turn our attention to what seems to matter about user psychology when it comes to virtually real experience.

Although research in this area is nascent, we have evidence for thinking that a user's position within the dissociation spectrum (i.e., how strong her tendency is relative to others to separate herself from the content of her experiences) is an important variable in terms of figuring out when and why simulated experiences are felt as virtually real. Additionally, their relative possession of some of the "Big Five" personality traits also seems to affect the degree to which a user can immerse themselves in mediated experiences and thus can affect how they will react to virtual experiences.

But first, a word of caution. Many of the claims I make in this chapter are empirically tenuous, at best. Extant research on virtual and especially augmented reality phenomenology is scant, and such research is important if we are to distinguish between experiences of presence and virtually real experiences. Additionally, the research that *is* available is subject to concerns over replication that we're all wise to keep in mind (Shrout & Rodgers 2018). As such, I'm only partially committed to the claims that follow.

My own understanding of current research, coupled with my experience of designing and running subjects through virtual reality simulations, leads me to conclude that "presence" and "virtually real" experiences are distinct, often overlapping, experiences. The elements of a simulation associated with these experiences that I lay out below should be considered a first pass at a theory of virtually real experience, and are subject to revision. We're in the early stages of building worlds and most of the ethical implications that follow can be separated from my specific views about perspectival fidelity or context-realism.

Context-realism

The most obvious place to begin an analysis of the elements of a virtually real experience is with those features of a simulation that deal directly with its narrative. Context-realism is a way of talking about how well a simulation's content – in other words how its physical rules, setting, non-player character (NPC) behaviors, overall narrative, and so on – cohere with a user's expectations of how real-world content is supposed to work. The more the physical and psychological rules of a simulated world cohere with a subject's expectations, the more context-real that simulation is for that user. This way of thinking about context-realism is not only dimensional (a simulation can be less context-real than some but more context-real than others), it's also subjective. In other words, one and the very same simulation can be more context-real for some users and less context-real for others depending on their background beliefs.

Before continuing, I would like to say a bit more about the distinction between virtual and augmented realities. Although these are currently seen as two different platforms (with distinct hardware, software, applications, and so on), the distinction between AR and VR is porous and can be downright fuzzy. In the future, as these technologies mature, it's more likely that they'll blend into one another. This can help explain why some futurists are already using the term "extended reality" to refer to the spectrum of technologies encompassed by the different forms of virtual reality (hmd, CAVE, etc.) and augmented reality technologies. At present, the major difference between VR and AR is a matter of degree: how much of our experience is simulated? Augmented reality devices like Google's Glass and Microsoft's Hololens don't aim to immerse users within the context of a completely simulated environment. Instead, they aim to add simulated content to a user's current environment (recall Luis' AR hauntings). VR simulations, on the other hand, tend to immerse users within entirely simulated spaces.[6]

Games like *Pokemon Go!* (Niantic 2016), for example, rely heavily on the user's physical location. The game itself adds onto our physical environment by placing simulated game elements (Pokemon, Gyms, etc.) into the physical environment. We can imagine games like this adding even more elements to existing environments. For example, imagine a future version of such a game that, while still keeping the basic structure of a user's physical world, applies graphical filters to that experience in order to make the world appear animated or monochrome or to give it a cyberpunk feel. Such a game seems to straddle the boundaries between AR and VR as they're currently conceived. As these technologies mature, it's far more likely that they will blend together instead of drift apart and that a single piece of hardware will allow users to determine how much of their experience consists of simulated content. All of this is to say that when I speak of perspectival fidelity and context-realism, my focus will be on VR simulations, but this is largely a historical artifact.

One of the most basic features of a simulation, VR, AR, or otherwise, is the set of rules that delimit the space of action allowed within it. Samaya's simulation allows her to fly around a virtual bridge as she works on drafting its engineering plans. Such a feature is incredibly useful for an architectural program and it's easy to see why a structural engineer might benefit from having this ability in a VR space. However, the very same feature almost certainly diminishes the degree of context-realism that the simulation has.

The physical rules of that simulation allow Samaya to do something (float around a structure) that she's not able to do in the real world (in part that's what makes the simulation so useful). In doing so, the simulation is likely to be experienced by Samaya as something less than real, even if the experience is one that she's highly present in. Additionally, the user interface (UI) built into the simulation is context-unreal in the sense that the tools and other elements of the UI float in space, awaiting commands and gestures from the user.[7] Each feature built into Samaya's simulation is a design choice. Such elements are, again, extremely useful for an engineering simulation but the very same elements diminish the degree to which that world works as our world does, reducing its context-realism. Assuming that the world doesn't contain other traditional real-world elements (time does not pass in it, diegetic sounds are missing so as to not distract from work, etc), then such features are also likely to detract from the simulation's context-realism (even if they simultaneously *add* to the usefulness of the simulation for its intended purpose). Similarly, we can imagine Giordan's game as containing features that diminish its context-realism.

Setting impacts a simulation's degree of context-realism. Giordan's game is set in a distant future with weapons and beings that he's not likely to believe exist in the real world (or, at least, that he's not likely to believe exist *right now*). Having an experience in such a setting, even a very photorealistic one, is unlikely to be treated as if it were really happening.[8] Such fantastic elements may make for a better fighting game but, like Samaya's floating tools, also make the simulation less context-real. Compare the setting of Giordan's simulation with Sam's simulation. Sam's therapeutic setting is, by design, intended to match up with user expectations of what the world, here and now, looks and sounds like. For that reason, contemporary settings are more context-real than science fiction futures or fantasy pasts. This helps in Luis' case too. Although his is an AR experience, the fact that his AR simulation uses his real-world surroundings increases its context-realism. That the elements of his hauntings are diegetic are why it seems so scary.[9] Similarly, we can see that Giordan killing the techno-zombie hordes in his game might make for good fun in ways that killing highly realistic present-day humans might not be. Context-unreal virtual agents are less likely to produce experiences that are treated as if they were when compared with more context-real virtual agents.

Another feature that's likely to diminish the context-realism of Giordan's simulation (and just about all violent video games) is his character's ability to dish out, and survive, impossible amounts of damage. Characters in games like these can survive falls that would kill a normal person and, further diminishing the context-realism of these simulations, they can heal instantaneously by walking over items or pausing the universe altogether and using a healing item (another set of element that's likely to detract from a simulation's context-realism). That such games also typically allow players many (sometimes infinite) ways of returning after death reduces the context-realism even more.

Consider also that such games are likely to contain other elements that often detract from its context-realism: enemies often disappear when they die or offer flashing powerups for the player to collect, the inhabitants of these virtual worlds often exhibit only a few stereotyped repetitive behaviors, and often such simulations include magic or supernatural elements. All of these detract from the context-realism of the simulation. Much as with Samaya's program, the features that make it a good instance of its type (architectural design simulation, first-person shooter, etc.) also work to diminish the simulation's context-realism.

Surprisingly, photorealistic environments don't appear to be a very important element of context-realism. Philosophers Jon Cogburn and Mark Silcox have reached a similar conclusion with respect to photorealism and presence: "the evidence so far does not support the contention that visual realism is an important contributory factor to presence" (Cogburn & Silcox 2014). Recall the study from the previous chapter which replicated Stanley Milgram's obedience experiments (Sanchez-Vives & Slater 2005; Slater et al. 2006). In that study, the virtual learner was only very crudely rendered and, despite this lack of NPC photorealism, at least some of Slater's subjects still responded to administering shocks to her in virtually real ways. Why is that?

What we learn from studies like this is that naturalistic settings (a mundane laboratory) and naturalistic behavior (pleading and realistic crying in the case of Slater's virtual learner) cohere with the psychological expectations that most people are likely to bring to an experience (real or virtual) and, in light of that coherence, are more likely to treat their experiences as virtually real. Similar results have been found across different VR simulations. For example, David-Paul Pertaub and his colleagues (2002) discovered that audiences in a virtual public speaking task reacted in a realistic, negative, way to giving an improvised speech to a (human but not photorealistic) crowd designed to realistically convey boredom and annoyance. Ryan McMahan and his colleagues (2012) conducted a more recent, and in many ways more telling, comparative study that supports the view I'm laying out here.

McMahan wanted to determine how behavioral realism (e.g., how accurately non-player characters, NPCs, in a virtual space act) and photorealism (how visually similar a simulation's graphics appear like our world) impact user experience in a VR simulation.[10] McMahan had his subjects play one level of a VR first-person shooter designed specifically for the study. There were four conditions in this study depending on what mix of behavioral realism (high/low) and photorealism (high/low) subjects were exposed to. In the behaviorally realistic condition, subjects themselves not only moved through virtual space in a more naturalistic fashion (using what they called a "human joystick" as opposed to a less realistic keyboard), but were confronted with more realistic behavior from their virtual opponents. According to McMahan and his colleagues,

> [i]n the high-interaction conditions, participants tended to move more using the human joystick technique than they moved with the keyboard technique in the low-interaction conditions. This difference in strategies resulted in participants taking less damage and being less accurate with high interaction fidelity while taking more damage and being more accurate with the low level of interaction fidelity.
>
> (McMahan et al. 2012)

These results are interesting, and telling. The less realistic a virtual environment, the greater a subject's accuracy when firing on their opponents (and the greater damage subjects would receive back from their virtual enemies in order to get that higher accuracy). The more realistic the environment, in terms of context-real elements like behavior, movement, and NPC strategy, the less accurate a subject became and the less damage they took while playing.

This should make sense to us. None of McMahan's subjects were trained soldiers and hence it's reasonable to see his study as telling us that naive and untrained people thrust into a realistic combat zone facing NPCs that acted like real people will be more likely to prioritize safety by cowering (in fear?!), firing more haphazardly at their enemies, and likely that they would take less damage than someone who is acting as they would if they were playing a video game like Giordan did when facing off against the techno-zombies. Subjects in McMahan's high realism simulation, in other words, treated their experiences as more real than those in the less realistic condition and their behavior is consistent with that conclusion.[11] While McMahan's CAVE-style VR system certainly created a sense of presence in his subjects, only subjects in the high fidelity condition seemed to treat their experiences in ways that look like realistic responses to threats.

Similarly, David Zendle, Daniel Kudenko, and Peter Cairns investigated the degree to which graphical realism and behavioral realism in simulated environments were associated with the activation of aggressive concepts (Zendle, Kudenko, & Cairns 2018). In one experiment, they "investigated whether making in-game enemies behave realistically via the use of ragdoll physics increased the activation of aggressive concepts in players" while in another they "looked at the effects of realistic nonplayer character (NPC) tactics" on aggression (21). Ragdoll physics, as the name implies, is a term used in simulation design to refer to a set of physics rules applied to avatars (all of an avatar's limbs are made to go limp and can each respond individually to forces applied on them and hence their bodies look, as the term implies, a bit like ragdolls).

In terms of context-realism, ragdoll physics, while an improvement over instances in which a defeated non-player character simply disappears, is still a less context-real way of simulating the behavior of deceased persons. That being said, consistent with McMahan et al.'s results, and with the view that context-realism plays an important role in the generation of virtually real experiences, Zendle, Kudenko, and Cairns (2018) found that while "the presence of ragdoll physics in a [violent video game] has little effect on the activation of aggression-related concepts" (25), that "[p]articipants who played the [first-person shooter] with realistic NPC tactics showed less activation of aggressive concepts" (28). These results also make a lot of sense. The more realistic the behaviors of a virtual person (as demonstrated by having NPCs use more human-like A.I. tactics against the player), the less likely a subject of such an experience is to behave like an invincible superhero (as they might with most games) and the more likely they are to act as they would in real life (i.e., with diminished aggression). Increasing a simulation's degree of context-realism, in this case via NPC behavior, increases the likelihood that a subject will treat that experience as if it were really happening.

At the highest degree of behavioral context-realism, we can expect virtual agents to demonstrate what some have called "moderate reasons-responsiveness" (Fischer & Ravizza 1998). A moderately reasons-responsive agent is one who acts intelligibly (in ways we understand) on the basis of the reasons available to her.[12] A virtual agent who appears to act in such ways will thus increase the context-realism of a simulation over one which contains virtual agents who only behave within a very narrow range of automated behaviors (demonstrating an insensitivity to even very powerful and obvious reasons). Virtual agents who realistically mimic user expressions, as humans often naturally do, can even trigger what's known as the *chameleon effect* in subjects. The chameleon effect refers to our tendency to perceive people more positively, as more prosocial, and as more deserving of help, when they mimic our behavior (Bailenson & Yee 2005).

As the preceding results suggest, behavioral realism appears to be a much more important element of context-realism than graphical realism. We'll see even more evidence of this when we look at experimental replication of Stanley Milgram's (1963) obedience study, gathered using virtual reality environments.

Giordan's techno-zombies are thus not especially context-real. This is the case not only because Giordan is unlikely to believe that techno-zombies are real but also because they don't act in realistic ways to being slaughtered. Assuming that Luis came into his AR experiences believing in ghosts, his reaction to the virtual haunting is context-real in ways that Giordan's experiences were not.

Similarly, Sam's virtual experiences, on the other hand, are highly context-real. The world represented in that simulation is, in many ways, structurally and physically similar to the world Sam experiences outside the simulation. The behavior of virtual agents in the simulation coheres with Sam's expectations of what real-life agents are supposed to act like. VRET simulations have proven so successful in part because they allow therapists to manipulate each simulation's degree of context-realism (and, as we'll see, a simulation's perspectival fidelity) to slowly empower patients to gain a sense of control over their triggers. Context-realism is only one of the elements of a virtually real experience, albeit an important and powerful one. Anyone who designs a simulation can exercise some control over that simulation's degree of context-realism. The same can be said about other, non-content, design features of a simulation. Let's turn to those next.

Perspectival Fidelity

Creating a simulation involves a series of design decisions not only about a simulation's narrative and not only about the simulation's physical rules (context-realism) but also about how that simulation will present itself to users. Perspectival fidelity is a way of referring to the design elements of a simulation that, while not really about its content, are tied to the choices built into the point-of-view created for users. Like context-realism, perspectival fidelity is dimensional. Simulation A might be more perspectivally faithful than simulation B while being less faithful than simulation C. Similarly, the more perspectivally faithful a simulation, the greater the likelihood of that simulation producing experiences that are treated as if they were real by a subject.

The contours of perspectival fidelity map out how information is presented to a subject. For example, a simulation designed to have a three dimensional first-person perspective and which uses a virtual reality head-mounted display (hmd) with a wide field of view is more faithful than one designed to have a third-person, two dimensional, or "God's eye" view (Roettl & Terlutter 2018). The reason why is obvious: our real-world experiences are all (or almost all) in the first-person. Similarly, a simulation that includes only diegetic sounds and soundtracks (sound and music produced by objects in the simulation instead of overlaid on top of it) is more perspectivally faithful than a simulation that includes non-diegetic sounds. With the exception of hallucinatory sounds, our own lives contain only diegetic sounds and soundtracks. Because almost all AR simulations currently in use overlap simulated content onto a user's actual perspective, AR applications can often be higher in perspectival fidelity "on the cheap." That is, AR applications that overlap information onto our experience can co-opt our perspective automatically without needing to design for it. This doesn't mean that AR simulations are always going to be more perspectivally faithful than VR simulations (e.g., *Pokemon Go!* is almost certainly less perspectivally faithful and context-real than VRET simulations) but it does mean that AR designers can get highly perspectivally faithful simulations more easily than VR designers.

Haptic feedback is, for most of us, another inescapable feature of being in the world. Picking up an apple, typing on a keyboard, walking, and driving all have distinctive real-world feedback that our bodies use to help us manipulate objects and make sense of our surroundings (Stone 2000; Williams 2014). When designing virtual and augmented reality simulations, there are many ways designers can choose to deliver haptic feedback, some more perspectively faithful than others. For example, a VR simulation where subjects move using a keyboard is less perspectively faithful than one where they are allowed to naturalistically move around a space. This is true for two reasons. The first, and most obvious, is that most of us physically move around real spaces to get where we're going and so simulations that mirror this kind of locomotion, sometimes referred to as simulations that use "natural mapping," more closely mirror our own experience:

> Research has shown that more naturally mapped control schemes often lead to greater feelings of presence, realism, and enjoyment among game players when compared to less naturally mapped control schemes.
>
> (Schmierbach, Limperos, & Woolley 2012).

Second, however, is that the haptic properties of a keyboard are radically different from the haptic properties of walking. Not only is physical feedback located in different places (the feet and legs vs. the fingers and hands), but our learned associations tied to these forms of feedback are different (e. g., moving around a space using a keyboard is more likely to be associated with gaming or working than it is for realistic movement, at least for most of us) (Williams 2014). Here too we might imagine AR applications having an easier time with perspectival fidelity in the sense that users move around the real world in order to move around virtual spaces and insofar as simulated objects are interacted with using our bodies (e.g., reaching into the refrigerator to grab an apple that has AR data overlaid onto it to give us nutrition information), haptics can also be had "on the cheap."

Similarly, we can imagine increasingly unfaithful ways of giving users haptic feedback for object interaction as well. Reaching out in a simulation and grasping an object with force-feedback gloves (gloves that provide a sense of pressure that can mimic an object's shape) is more faithful than reaching out with a control wand that vibrates when it collides with an object. Vibratory feedback like this is common in the current generation of commercial VR. Such vibrations are more faithful than a simulation that doesn't give users any feedback at all or that provides non-diegetic feedback (e.g., simulations that use a tone or verbal warning to let users know when the user is near an object) (Schmierbach, Limperos, & Woolley 2012).

Similarly, simulations like Samaya's and Giordan's, both of which overlay meta-information onto their field of view, are less perspectively faithful than simulations that don't do this. In Samaya's case her overlay included architectural and engineering tools. In Giordan's case it may include an in-game map, score and health markers, or a targeting reticule, ammo, etc. The degree to which meta-informational elements diminish perspectival fidelity may change as augmented reality devices make these forms of meta-information more common parts of our lived experiences. For the moment, however, very few of us have day-to-day experiences that contain this sort of augmented reality meta-information.

Notice, however, that in all of these cases, perspectively fidelity is, like context-realism, relative to individual subjects. Those more used to assisted movement (e.g., via a joystick enabled motorized wheelchair) will experience a simulation that uses a joystick for locomotion as *more* perspectivally faithful than someone whose experience with locomotion is limited to walking and running. A simulation that presents users with neurotypical color vision, is set at average height and so on, is more perspectivally faithful for those populations than a simulation that mimics seeing in the infrared or that gives users an ant's-eye view of our world.

Hardware features, beyond controller/input haptics, also play a role in perspectival fidelity. For example, nearly all modern virtual reality platforms utilize an hmd to deliver visual and auditory input to users. The properties of the screens inside those hmds (refresh rates, field of view, display resolution, etc.) can affect both the context-realism and perspectival fidelity of any simulation run through the hardware. The low refresh rates of early hmds, for example, made those early VR systems uncomfortable and nauseating to use. An hmd's capacity to generate stereoscopic sound is also tied to its perspectival fidelity for neurotypical persons. The weight of the hmd itself can diminish perspectival fidelity if it begins to intrude on a subject's experience (in this sense it can also diminish "being there" presence) (Fox, Arena, & Bailenson 2009; Sanchez-Vives & Slater 2005).

Some features of a simulation and its hardware affect a simulation's context-realism and perspectival fidelity. Non-diegetic visual or auditory cues, flashing icons, and shifts in first- and third-person perspective all diminish, to varying degrees, both a simulation's context-realism and perspectival fidelity. They diminish these features because our experiences don't contain such cues, at least for most of us, and hence the presence of these elements within a simulation serves as a reminder that the experiences are taking place within a simulated environment. Such reminders thus diminish the likelihood that a subject will treat experiences in such simulations as if they were real. Additionally, whether a designer chooses to represent moral dilemmas as personal or impersonal (Greene et al. 2001) can affect not only how users respond to the dilemmas but can affect other design decisions that impact perspectival fidelity and context-realism.[13]

Designers of virtual reality simulations can exercise a great deal of control over the degree to which their simulations are likely to generate virtually real experiences. This control is an essential element of the ethical guidelines I offer in the following chapters for how to design ethical VR and AR content. However, although designers of VR and AR simulations can exercise a lot of control over how subjects are likely to experience their simulations, a subject's psychological traits also play an important role in this account. Some subjects, we should think, might be more apt to treat their simulated experiences as if they were real than other subjects.

Subject-psychology

Psychologists, philosophers, therapists, and game designers are all in the business of creating virtual worlds. As we'll see in the next chapter, some of these groups have already run into what could (and should) have been preventable ethical problems. Because those of us who design virtual worlds can control elements of our simulations, such as their degrees of perspectival fidelity and context-realism, we can exercise a lot of control over, and are thus morally responsible for, how our simulations affect users. One of the aims of this book is to make clearer what the emerging ethical parameters are for virtual and augmented reality simulations, and to offer guidelines for how we ought to design virtual worlds (those worlds which arrive at the best balance between potential virtual harms and ethically permissible ends).

However, though designers have significant control over many aspects of their virtual worlds, I've neglected to say much about how individual psychology might make some people more or less likely to experience simulations as virtually real. Virtually real experiences are *experiences*, meaning that they are had by people, and people, though they may share many features, are idiosyncratic beings. Although we can speak generally about the features of a simulation that can make it more or less likely to generate virtually real experiences, it's important to examine the elements of individual psychology that can dispose someone to experience perspectivally faithful and context-real simulations, even those with poor graphical realism, as virtually real.

I'll begin by noting that the study of user psychology and its relationship to virtual reality is a growing field but its contours are still being defined. Thus, the evidence we have about the psychological features relevant to these questions is scant, and subject to all of the sorts of replication issues already discussed. What follows should be taken as even more preliminary and subject to revision than everything else I've written in this chapter. I expect that this field of research will grow immensely in the next decade but, for now, we can make predictions based on extant data.

That being said, there do seem to be important elements of a user's psychology that can influence the degree to which individual users treat a virtual experience as if it were real. The evidence available suggests that a person's relative possession of some of the "Big Five" personality traits, especially the traits known as openness to experience, neuroticism, and extraversion can affect their likelihood of immersing themselves in fictive experience and thus can influence their response to virtual worlds.

Additionally, a subject's tendency to dissociation, as measured by something like the Dissociative Experiences Scale, especially their tendency toward normative dissociation (Seligmann & Kirmayer 2008), can impact how likely that subject is, not only to experience themselves as "present" in a simulation, but also to have virtually real experiences within them. I'll examine this evidence too.

When psychologists talk about the "Big Five," they're referring to sets of large-scale (what philosophers might call "second-order") dispositions that tend to have all sorts of specific effects on how we think and feel in certain situations and about what sorts of things we value (de Raad & Perugini 2002). These dispositions can affect and regulate many other more specific behavioral, emotional, and cognitive traits. The Big Five personality traits include:

1   Openness to experience: whether someone enjoys variety, is imaginative, interested in new ideas. Higher scores indicate the degree to which someone enjoys and is open to these experiences
2   Conscientiousness: whether someone is reliable, careful, hard working, and has a strong conscience. Higher scores indicate greater reliability
3   Neuroticism: the degree of emotional stability exhibited by a person. High scorers are less emotionally stable (they respond with greater emotionality)
4   Extraversion: the degree to which someone's attention is directed toward others. Higher scores indicate that someone is social, outgoing, and assertive
5   Agreeableness: how trusting, cooperative, and accommodating someone is. Higher scores indicate greater degrees of these traits

(Soldz & Vaillant 1999)

David Weibel and his colleagues conducted a study to determine whether any of the Big Five personality traits played a significant role in explaining why it is that some people seem to be able to more easily immerse themselves in fictive experiences than others (Weibel, Wissmath, & Mast 2010). Weibel and his colleagues discovered that, although the Big Five are important in many ways, not all of them affected how people experience simulated worlds. According to them, their

> findings show that people who score high on the Big Five dimensions openness to experience, extraversion, and neuroticism more likely immerse in media than do individuals scoring low on those dimensions. Therefore, we can conclude that personality traits determine how people experience media exposure and whether someone easily experiences sensations of presence while being confronted with mediated environments.

(Weibel, Wissmath, & Mast 2010)

This study has several limitations, and we should be cautious about how much we generalize from it when we talk about the psychology of virtual experiences. For instance, the study is focused very generally on all sorts of mediated experiences including reading, watching movies, and playing games. The traits relevant to presence and virtually real experience may only be weakly related to the traits that determine whether someone is an especially imaginative reader, for instance. Second, the study identifies personality traits entirely using people's responses to questionnaires about their own dispositions. Given the possibility of self-deception and imperfect self-knowledge, such data can misrepresent the degree to which people really do possess the personality traits in question. Despite these limitations, Weibel believed that their results suggest that the conclusion that

> virtual reality (VR) technologies can elicit stronger experiences of presence than conventional media such as TV might be particularly true for people with high openness scores. Thus, it is likely that the confrontation with a VR technology like a CAVE satisfies the curiosity of such individuals and therefore causes strong immersive experiences. We can therefore conclude that the influence of new media, which were designed to absorb and immerse their users – such as e-learning or cybertherapy settings – might be most effective for users scoring high on openness to experience. In the future, personality traits can play a more important role as they influence how media exposure is received by users.
>
> (Weibel, Wissmath, & Mast 2010)

Tentatively then, we can conclude that the degree to which a specific individual possesses personality traits like openness to experience, neuroticism, and extraversion can impact the likelihood of that individual experiencing a perspectively faithful and context-real simulation as virtually real. Users who are more open to experience, more extraverted, and who score higher on neuroticism (i.e., who tend to respond to experiences with greater emotionality) are not only more likely to seek out VR (and probably also AR) experiences, they're also more likely to treat their experiences as virtually real.

Earlier in the chapter, I spoke of presence as a basic psychological capacity to feel located in one's experiences. Although not identical, the concept of presence is closely related to another dimensional psychological capacity called dissociation. Evidence suggests that one's tendency toward dissociation also plays a role in determining a person's likelihood of experiencing simulated events as if they were real.

Dissociation is a complex term, about which there is significant controversy, especially between psychiatric, anthropological, and philosophical models (Seligmann & Kirmayer 2008; Lynn 2005; Hollan 2000; Snodgrass 2004). One reason for this controversy stems from how complex the phenomena associated with dissociation are. Dissociation is often discussed as "involving functional alterations of memory, perception and identity as well as the psychophysiological processes presumed to underlie these phenomena" (Seligmann & Kirmayer 2008; Spiegel & Cardeña 1991). For our purposes, let's refer to dissociation as pointing to the range of possible ways in which people can disentangle themselves from their experiences. Although dissociation is perhaps best known in terms of its pathological and extreme forms (Dissociative Identity Disorder, Depersonalization Disorder, etc.), to some degree, dissociation is a part of normal human experience and is even sometimes the kind of experience that people seek out and even come to value.

For example, the feeling of being in a "flow" state, one where we lose track of our experience of time while engrossed in an activity, is a dissociative state (Csikszentmihalyi 1996). Similarly, daydreaming and highway hypnosis (i.e., losing a sense of the passage of time while driving) are states where someone dissociates from their environment, their location, and their sense of the passage of time. Psychologists typically understand dissociation as a dimensional feature and measure it using something called the "Dissociative Experiences Scale" (DES) (van IJzendoorn & Schuengel 1996) or the "Cambridge Depersonalization Scale" (Aardema, O'Connor, Côté, & Taillon 2010). There is some evidence that "[e]pidemiological studies indicate that the life-time prevalence of [depersonalization and derealization] is 34% to 70% in nonclinical populations, suggesting that some level of dissociative experience is a normal phenomenon" (Aardema, O'Connor, Côté, & Taillon 2010, 429).

All of us have a degree of dissociative tendency, that is, we all have a relative degree of how likely we are to have experiences from which we feel depersonalized (that we're not real or not really involved in the experience) or derealized (that the world around is not real). Put this way, it's clearer that dissociative tendency probably plays an important role in determining the likelihood of specific individuals being "present" in simulated experiences or of experiencing these events as if they were real (as virtually real). Very little data has been gathered on the relationship between presence, dissociation, and virtual experience.

In particular, studies on the relationship between dissociative tendencies and presence are vanishingly scarce. Frederick Aardema and his colleagues (2010) provide us with one data-point that's roughly in this neighborhood. However, in their case, they were interested in the opposite relationship to the one we're interested in. Aardema et al. were interested in examining how exposing subjects to virtual reality experiences would affect their likelihood of dissociating from their real-life experiences.

Aardema et al.'s results are fascinating and point toward there being a relationship between user psychology, especially a user's degree of dissociation, and virtual experience. Aardema claimed to find that "[i]mmediately after exposure to the VR environment ... participants did not reach the same level of presence in objective reality as they had before"; further, their "[r]esults showed that change in symptoms of [depersonalization and derealization] due to VR exposure … was most strongly related to pre-existing levels of [depersonalization and derealization]" (Aardema, O'Connor, Côté, & Taillon 2010, 433). The implication from these findings is that a subject's pre-existing levels of depersonalization and derealization played an important role in determining the degree to which the subject would go on to dissociate from real-life experiences following VR exposure.

Other studies have followed Aardema et al. in focusing their investigation on whether VR can *cause* dissociation toward real-life experiences but not, unfortunately, on whether a person's prior tendency to dissociate can impact how they experience virtual experiences. Dalena van Heugten–van der Kloet and her colleagues demonstrated that out-of-body VR experiences could create dissociation in subjects after they emerged from virtual spaces (2018). Echoing Aardema's data, van Heugten–van der Kloet's team note that their "findings highlight a significant increase in acute dissociation after VR exposure" to an out-of-body VR simulation (van Heugten–van der Kloet et al. 2018, 350).

That user psychology is an important element of presence and virtually real experiences isn't surprising. Even when discussing context-realism and perspectival fidelity, it was important for me to mention individual psychology in order to make clear that both concepts rely on discovering whether a virtual world cohered with a particular subject's expectations about how the real-world looks, feels, and behaves from the user's own point of view. To these features, we can now add the degree to which a subject is open to experience, narcissistic, and agreeable, while their tendency to dissociation must also feature more heavily in an analysis of both presence and virtually real experience.

Although I've mentioned that VR and AR designers can control their simulations' degree of perspectival fidelity and context-realism but that they can't control user psychology, this isn't always true. For example, in any experimental context, like many of the virtual reality human subjects research I've already mentioned, subjects can be screened prior to formally entering virtual spaces. This should be done in order to minimize potential harms that might come from virtually real experiences (more on this in the next chapter).

The picture emerging from these data is that we must move beyond studying virtual experiences by reference to the concept of presence alone. Presence, especially the "being there" sense of presence, is not rich enough to make sense of the variety of ways users can relate to their virtual experiences. "Being there" and "doing there" senses of presence tell us part of the story about how a subject relates to their body and to virtual objects. However, if we want to make sense of the differences between Samaya and Giordan's experiences on the one hand and Luis and Sam's experiences on the other, then we need to invoke a concept like virtually real experience to make sense of these differences and their variety.

Furthermore, virtually real experiences themselves appear to be composed of at least three overlapping dimensions: a simulation's degree of context-realism, its degree of perspectival fidelity, and individual user psychology. Figure 3.1 represents this emerging picture. If we build simulations high in perspectival fidelity and context-realism then we increase the likelihood that such simulations will be experienced as virtually real by their users. When users with the right mix of psychological dispositions respond to such simulations, the likelihood of the subject having a virtually real experience is higher still.

Because two of these elements are under the direct control of those of us who develop and design virtual environments, they bear responsibility for how their users respond to these simulations.

Conclusions

I've argued that virtually real experiences are distinct, in important ways, from those experiences that psychologists refer to as containing "presence." While the illusion of "being there" and even of "doing there" are important elements of virtually real experiences, I've argued that, by themselves, these concepts do not consistently generate experiences that subjects are likely to treat as if they were real (i.e., as virtually real). Virtually real experiences are more likely to occur in subjects with dissociative tendencies placed into simulations high in perspectival fidelity and context-realism. That virtually

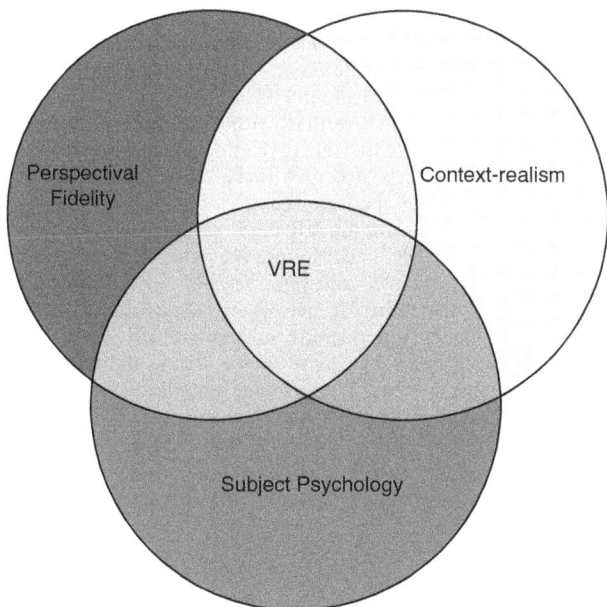

Figure 3.1 Virtually Real Experiences (VREs)

real experiences are possible creates a set of ethical issues that are unique to VR and AR spaces. I turn my attention to one set of these ethical issues, those pertaining to experimental ethics, in the next chapter. Following that, I'll examine other ethical issues arising for commercial and political uses of such simulations.

Notes

1 Questions are interesting, in the sense I'm referring to here, when they're unique to the medium. We can always assess a simulation's ethics (virtual, theatrical, literary, filmic, imaginary) in terms of its content (veridicality, bias, utility, etc.) but only simulations capable of producing what I'll be calling "virtually real experiences" give rise to unique issues with VR technology.

2 Although SafeSpace is made-up software, virtual reality exposure therapy is a very real, and very promising, application of VR technology (Opriş et al. 2012). I'll say much more about VRET, and why we can learn a lot about virtual experiences from studying it, later on in this chapter.

3 I first described these elements in my (2017) "Empathy and the limits of thought experiments" but have since (2018a; 2018b) refined the concepts considerably and the presentation of virtually real experience, and its components, laid out in this chapter represent what I consider an important improvement over these earlier accounts, especially because they now center-subject psychology more prominently and offer a specific account of the AR/VR spectrum and its connection to virtually real experiences.

4 Two exceptions are notable. First, age-gating VR/AR applications is straightforward and is something already done for a wide variety of activities and media. I'll say more about the effects of age on VR/AR ethics in Chapter 6. Second, experimentalists and therapists can (and have a moral responsibility to) screen their subjects to avoid exposing especially at-risk subjects (those that possess psychological dispositions that make them especially likely to have virtually real experiences while using VR/AR) to unintended harm.

5 It's worth pausing to point out here that neither of the three senses of presence described here make reference to the psychological dispositions of particular subjects. Psychologists, as these accounts make clear, tend to focus on subject phenomenology when writing about presence.

6 The fact that even current generation VR hmds contain outward facing cameras, however, tells us that even VR hmds can use the external world in the same way as AR systems do. In principle, someone wearing a VR hmd could use such cameras to run AR applications. It may even be possible to replicate classic thought experiments like "inverted spectrum" experiments using such a setup.

7 Interestingly, I think elements like these won't always diminish a simulation's degree of context-realism. Why? The growing popularity of augmented reality (AR) interfaces will soon make tools like these a common feature of our real-world experience and hence VR simulations that include these elements will be experienced as more context-real than they are today. This is yet another way in which context-realism is dependent on user psychology and beliefs about the world and how it works.

8 As I said a moment ago, there's an element of subjectivity here. Perhaps Giordan really and truly believes that he's a time traveller from the future and that the game is an accurate representation of what his life was like before visiting us. Although far-fetched, if Giordan has had such genuine beliefs then the simulation is going to be much more context-real for Giordan than it is for you or me.

9 Here too context-realism is relative to user beliefs. If Luis were an ardent, skeptical, atheist empiricist, we can imagine him treating the haunting simulation in much the same way that Giordan treats killing techno-zombies.

10 McMahan refers to behavioral fidelity as "interaction fidelity" and photo-realism as "display fidelity." McMahan also manipulated the input devices that his subjects used for locomotion in their virtual space. How users move about virtual spaces impacts both the simulation's context-realism (imagine driving with a keyboard instead of a steering wheel) but also its perspectival fidelity. I'll save discussion of haptics and locomotion until a later section of this chapter.

11 Although McMahan's group collected significant behavioral data during the study, they sadly did not collect emotional data during the study so it's impossible for us to tell whether they experienced any sort of subjective virtually real trauma. The group did, however, make decisions that were aimed at diminishing the simulation's context-realism (i.e., removing blood) even from the highest fidelity versions of the simulation. It may be that such decisions rendered the simulation low enough in context-realism to avoid causing significant virtual trauma. I'll say more about virtual trauma in the next chapter.

12 Fischer and Ravizza believe that moderate reasons-responsiveness (MRR) is a feature of mental mechanisms, not agents, and hence an agent can act on an MRR mechanism and be responsible in one case but not in others. This is an important distinction, one which I ignore in this chapter for the sake of simplicity in speaking about virtual agents. It's entirely possible, at the extremes of virtual agency, that they are essentially just like human agents and hence sometimes act on MRR mechanisms and sometimes not. For my purposes, I assume that no current (or near-future) virtual agents are actually acting on MRR mechanisms. What matters, for context-realism, is that they give users the *appearance* of being MRR.

13 For example, Foot's classic (1978) trolley problem is impersonal (in the technical sense in which you are not directly making contact with the people your actions affect), while Judith Thomson's (1985) emergency room variant of the trolley problem is personal (in the sense in which one must physically harvest a person's organs in order to save others). There is some evidence (Greene et al. 2001; Greene 2009) that we experience personal/impersonal moral dilemmas differently.

References

Aardema, F., O'Connor, K., Côté, S., & Taillon, A. 2010. Virtual reality induces dissociation and lowers sense of presence in objective reality. *Cyberpsychology, Behavior, and Social Networking*, 13 (4), 429–435.

Ahn, S.J., Bostick, J., Ogle, E., Nowak, K., McGillicuddy, K., & Bailenson, J.N. 2016. Experiencing nature: Embodying animals in immersive virtual environments increases inclusion of nature in self and involvement with nature. *Journal of Computer-Mediated Communication*, doi:10.1111/jcc4.12173.

Bailenson, J.N., & Yee, N. 2005. Digital chameleons: Automatic assimilation of nonverbal gestures in immersive virtual environments. *Psychological Science*, 16 (10), 814–819.

Cogburn, J., & Silcox, M. 2014. Against brain-in-a-vatism: On the value of virtual reality. *Philosophy and Technology*, 27 (4), 561–579.

Csikszentmihalyi, M. 1996. *Creativity: Flow and the psychology of discovery and invention*. New York: Harper Perennial.

Cummings, J., & Bailenson, J. 2016. How immersive is enough? A meta-analysis of the effect of immersive technology on user presence. *Media Psychology*, 19 (2), 272–309.

de Raad, B., & Perugini, M. (Eds.) 2002. *Big five assessment*. Ashland OH: Hogrefe & Hubner.

Fischer, J.M. & Ravizza, M. 1998. *Responsibility and control: A theory of moral responsibility*. Cambridge: Cambridge University Press.

Foot, P. 1978. The problem of abortion and the doctrine of double effect. In *Virtues and vices and other essays in moral philosophy*. Oxford: Clarendon Press.

Fox, J., Arena, D., & Bailenson, J. 2009. Virtual reality: A survival guide for the social scientist. *Journal of Media Psychology*, 21 (3), 95–113.

Greene, J.D. 2009. Dual-process morality and the personal/ impersonal distinction: A reply to McGuire, Langdon, Coltheart, and Mackenzie. *Journal of Experimental Social Psychology*, 45 (3), 581–584.

Greene, J.D., Sommerville, R.B., Nystrom, L.E., Darley, J.M., & Cohen, J.D. 2001. An fMRI investigation of emotional engagement in moral judgment. *Science*, 293 (5537), 2105–2108.

Guterstam, A., Gentile, G., & Henrik Ehrsson, H. 2013. The invisible hand illusion: Multisensory integration leads to the embodiment of a discrete volume of empty space. *Journal of Cognitive Neuroscience*, 25 (7), 1078–1099.

Hegedüs, G., Darnai, G., Szolcsányi, T., Feldmann, Á., Janszky, J., & Kállai, J. 2014. The rubber hand illusion increases heat pain threshold. *European Journal of Pain*, 18, 1173–1181.

Heidegger, M. 1927/2011. *Being and time*. Macquarrie, J. & Robinson, E. (translators). New York: Harper & Row.

Hollan, D. 2000. Constructivist models of mind, contemporary psychoanalysis, and the development of culture theory. *American Anthropologist*, 102 (3), 538–550.

Horrigan-Kelly, M., Millar, M., & Dowling, M. 2016. Understanding the key tenets of Heidegger's philosophy for interpretive phenomenological research. *International Journal of Qualitative Methods*, 15 (1). https://doi.org/10.1177/1609406916680634

Lynn, C. 2005. Adaptive and maladaptive dissociation: An epidemiological and anthropological comparison and proposition for an expanded dissociation model. *Anthropology of Consciousness*, 16 (2), 16–50.

McMahan, R.P., Bowman, D.A., Zielinski, D.J., & Brady, R.B. 2012. Evaluating display fidelity and interaction fidelity in a virtual reality game. *IEEE Transactions on Visualization and Computer Graphics*, 18 (4).

Milgram, S. 1963. Behavioral study of obedience. *Journal of Abnormal and Social Psychology*, 67, 371–378.

Niantic Labs. 2016. *Pokemon Go!* VR simulation. https://pokemongo.nianticlabs.com/en/

Opriş, D., Pintea, S., García-Palacios, A., Botella, C., Szamosközi, Ş., & David, D. 2012. Virtual reality exposure therapy in anxiety disorders: a quantitative meta-analysis. *Depression and Anxiety*, 29 (2), 85–93.

Pertaub, D.P., Slater, M., & Barker, C. 2002. An experiment on public speaking anxiety in response to three different types of virtual audience. *Presence Teleoperators and Virtual Environments*, 11 (1), 68–78.

Ramirez, E. 2017. Empathy and the limits of thought experiments. *Metaphilosophy*, 48 (4), 504–526.

Ramirez, E. 2018a. Ecological and ethical issues in virtual reality research: A call for increased scrutiny. *Philosophical Psychology*, 32 (2), 211–233.

Ramirez, E., & LaBarge, S. 2018b. Real moral problems in the use of virtual reality. *Ethics and Information Technology.* https://doi.org/10.1007/s10676-018-9473-5

Roettl, J., & Terlutter, R. 2018. The same video game in 2D, 3D or virtual reality: How does technology impact game evaluation and brand placements? *PLoS ONE*, 13 (7), e0200724.

Sanchez-Vives, M.V., & Slater, M. 2005. From presence to consciousness through virtual reality. *Nature Reviews: Neuroscience*, 6, 332–339.

Schmierbach, M., Limperos, A.M., & Woolley, J.K. 2012. Feeling the need for (personalized) speed: How natural controls and customization contribute to enjoyment of a racing game through enhanced immersion. *Cyberpsychology, Behavior, and Social Networking*, 15 (7), 364–369.

Seligmann, R., & Kirmayer, L.J. 2008. Dissociative experience and cultural neuroscience: Narrative, metaphor and mechanism. *Culture, Medicine, and Psychiatry*, 32 (1), 31–64.

Shrout, P., & Rodgers, J.L. 2018. Psychology, science, and knowledge construction: Broadening perspectives from the replication crisis. *Annual Review of Psychology*, 69, 487–510.

Slater, M., Antley, A., Davison, D., Swapp, D., Guger, C., Barker, C., Pistrang, N., & Sanchez-Vives, M.V. 2006. A virtual reprise of the Stanley Milgram obedience experiments. *PLoS ONE*, 1, e39.

Snodgrass, M. 2004. The dissociation paradigm and its discontents: How can unconscious perception or memory be inferred? *Consciousness and Cognition*, 13 (1), 107–116.

Soldz, S., & Vaillant, G.E. 1999. The Big Five personality traits and the life course: A 45-year longitudinal study. *Journal of Research in Personality*, 33, 208–232.

Spiegel, D., & Cardeña, E. 1991. Disintegrated experience: The dissociative disorders revisited. *Journal of Abnormal Psychology*, 100 (3), 366–378.

Stone, R.J. 2000. Haptic feedback: A brief history from telepresence to virtual reality. In Brewster, S. & Murray-Smith, R. (Eds.), *Haptic human–computer interaction.* New York: Springer, 1–16.

Thomson, J. 1985. The trolley problem. *Yale Law Journal*, 94 (6), 1395–1415.

van Heugten-van der Kloet, D., Cosgrave, J., van Rheede, J., & Hicks, S. 2018. Out-of-body experience in virtual reality induces acute dissociation. *Psychology of Consciousness: Theory, Research, and Practice*, 5 (4), 346–357.

van IJzendoorn, M.H., & Schuengel, C. 1996. The measurement of dissociation in normal and clinical populations: Meta-analytic validation of the Dissociative Experiences Scale (DES). *Clinical Psychology Review*, 16 (5), 365–382.

Weibel, D., Wissmath, B., & Mast, F.W. 2010. Immersion in mediated environments: The role of personality traits. *Cyberpsychology, Behavior, and Social Networking*, 13 (3), 251–256.

Williams, K.D. 2014. The effects of dissociation, game controllers, and 3D versus 2D on presence and enjoyment. *Computers in Human Behavior*, 38, 142–150.

Zendle, D., Kudenko, D., & Cairns, P. 2018. Behavioral realism and the activation of aggressive concepts in violent video games. *Entertainment Computing*, 24, 21–29.

4   Virtual Experience, Real Harm

I've been sketching out a way for us to think about the relationship between moral psychology, the nature of VR/AR technology, and philosophy of mind. We've explored not only empathy and the limits of thought experiments but also looked for a framework to make sense of the psychology of simulated experiences. Drawing on work within the philosophy of mind and psychological research on presence, we've pieced together different elements that seem especially important for anyone who wants to theorize, design, or use virtual reality technologies.

When simulated worlds high in context-realism and perspectival fidelity are experienced by someone with the right mix of psychological dispositions, virtually real experiences often result and their likelihood increases as these relationships increase. In some cases, those experiences can be fun, even therapeutic. Walking around the surface of Mars or virtually visiting places around the world that would be hard for you to get to can be uniquely entertaining. Virtually experiencing historical events can be genuinely educational. Other times, however ... things can go less well. While VR and AR technologies, and the virtually real experiences they can connect us to, represent a significant advance over the Thought Experiment Paradigm, these experiences come with their own risks. Simulations likely to create virtually real experiences can be unethical to design and they can hurt us in different ways. I'll turn my attention to these situations now.

Over the next few chapters I'll be discussing the ethics of VR and AR simulations, especially when and why it might be wrong to build simulated worlds in specific ways or for specific purposes. In this chapter I'll focus on the harm to users that virtually real experiences can cause. Subjects in VR experiments, consumers in the VR and AR marketplace, and students in VR or AR classrooms all might be (wittingly or unwittingly) harmed by poorly designed simulations that don't pay enough attention to the way that user psychology, context-realism, and perspectival fidelity interact.

In the next chapter I'll look at how VR simulations designed to enhance empathy may end up promoting intersectional injustice. In a still later chapter I'll consider other harms that VR and AR might cause to users that go beyond subjective trauma. It might be possible, in other words, for

DOI: 10.4324/9781003042228-4

someone to be harmed by a simulation even if they don't have any negative experiences in them. VR and AR simulations can manipulate us, and they can change us in ways that go against our own best interests. An ethics for these new technologies needs to be sensitive to the special ways that they can have these effects on us. Unless we're careful about how we build our worlds, VR and AR simulations may make us into (morally) worse people, so we'd better be careful!

To get us started, I want to start by exploring an especially famous case of experimental ethics gone awry.[1] It might seem strange to begin with a historical, decidedly low-tech, case but I'll argue that this example is not only instructive, it's prophetic. The following are real transcripts from a series of experiments carried out in the 1960s by a psychologist named Stanley Milgram, who, at the time, was conducting experiments at Harvard University. These studies are so important to the history of experimental ethics that it's likely that if you've heard of Milgram before, it's probably in connection with these experiments. I briefly discussed Milgram's research in earlier chapters but now we must confront this work directly to make better sense of not only how psychologists in the 20th century came to create stronger codes of experimental ethics (especially experiments involving human subjects) but also to help us be more proactive about the ethics of experiments in the 21st century using virtual worlds.

Milgram was interested in learning more about when, and to what extent, people were willing to obey authority. Coming off of the horrors of the Second World War, the question of how seemingly normal people could willingly participate in atrocities like the Holocaust was a matter of urgent research.[2] The experiments that Milgram conducted to answer this question have, over the following 60 years, become an important model for people interested in studying the effects of authority on obedience, but they're also required reading in any seminar on the ethics of experimenting on human subjects. In the most famous version of these experiments, Milgram invited his subjects to his lab under the false pretense of taking part in a study about learning and punishment.

Each of Milgram's subjects would be seated at a table on which sat a metal box containing a row of switches. Each switch was marked in ascending 15 volt increments starting with 15 volts and climbing all the way up to 450 volts (e.g., 15 volts, 30 volts, 45 volts). The switches were also marked with written ratings so that Milgram's subjects would have a sense of their severity, including ratings such as "moderate," "strong," "danger – severe shock," and, at the extreme end, "XXX." Subjects were instructed to read a series of (seemingly) random words out loud and then they were supposed to wait for a response from another subject they were told had been randomly assigned to become the "learner."[3] If the learner responded with the "wrong" answer (i.e., if they didn't repeat the word given to the subject as the correct response to the word pairing), then the subject was told that it was their job to shock the learner as part of their research on learning and punishment.

For every incorrect answer the learner gave, subjects were told to increase the severity of the shock by 15 volts. In the most famous version of the study, Milgram himself would be in the room, hovering over his subjects, throughout the experiment. If a subject ever refused to go on with the experiment or if they seemed unsure of what to do, Milgram's role was to politely but firmly tell them that they must continue. This was designed to be a stressful situation. Consider one's subject's response to giving a strong shock to the learner:

[SUBJECT]: eh … My nerves aren't very good mac. I'm sorry. I don't wanna give this fella any more.
[EXPERIMENTER]: Please continue.
[SUBJECT]: I'd much rather have you take this check back and just let me go home.
[EXPERIMENTER]: It's absolutely essential that you continue. Please go on.
(Hollander & Maynard 2016)

Many of Milgram's subjects responded like this. Intentionally hurting someone, even in the context of a study and even with supposed permission from the experimenter in the room with you, is hard for us to do. In reality, the learners in the experiment weren't actually being shocked. The switches were not connected to anything. The learners, whom subjects were told were fellow subjects randomly assigned to the role, were actually actors who were in on the study and not in danger. They were responding with pre-recorded messages designed to express more and more urgent cries of pain. They were instructed, at higher shock levels, to complain about their hearts, to ask to leave the experiment, and eventually to simply stop reacting at all (as if they were passed out or possibly dead).

[SUBJECT]: You're getting a hundred and ninety vive [sic] – ninety five volts. (flips shock switch)
[LEARNER]: (Yelps in pain) LET ME OUT OF HERE. LET ME OUT OF HERE! MY HEART'S BOTHERING ME. LET ME OUT OF HERE! YOU HAVE NO RIGHT TO KEEP ME HERE! LET ME OUT! LET ME OUT OF HERE! LET ME OUT!
[EXPERIMENTER]: Go on please continue.
[LEARNER]: GET ME OUT OF] HERE! MY HEART'S BOTHERING ME! LET ME OUT! LET ME OUT!
[SUBJECT]: I – I don't wanna be responsible for anything happening to him
(Hollander & Maynard, 2016)

At 405 volts (a severity that was labeled as "Danger: Severe Shock" on the shock panel subjects were using throughout the experiment), one subject expressed serious reservations about continuing with the experiment and tried to stop their involvement:

[SUBJECT]: I'm sorry I can't continue. (breathless voice)
[EXPERIMENTER]: The experiment requires that you go on. Please continue.
[SUBJECT]: But I don't want it on my conscience ... Can't you even check
    him first sir?
[EXPERIMENTER]: (clears throat)
[SUBJECT]: To be sure that he's alright?

(Hollander & Maynard, 2016)

Stanley Milgram's results were groundbreaking, though perhaps not surprising given Milgram's initial desire to better understand the Holocaust. All of the subjects in Milgram's initial study delivered shocks of at least 150 volts and two thirds were able to bring themselves to deliver *all* of the shocks (including the ones labeled "XXX")! However, as the dialogue above demonstrates, many of Milgram's subjects did this under extreme stress and psychological trauma.

Driven in part by the reactions his subjects had, Milgram's experiments set off a long and still ongoing conversation about the ethics of human experimentation and what kind of safeguards should be put in place to protect people from being unethically harmed in a study.[4] One of Milgram's subjects said that they

> wouldn't want to do another experiment like that again for any amount of money. I'm still sorry I went to do it. It took me a couple of weeks before I was able to forget about it. I don't think its [sic] right to put someone through such a nervous tension.
>
> ("Reaction of subjects" 1962)[5]

Because so many of Milgram's subjects experienced unpleasant emotions, stress, even trauma because of their participation in the study, and because they didn't benefit much from participating (unlike someone taking an experimental but risky drug, for example), Milgram's experiment could never be replicated today.[6] Following experiments like these, scientists began to take more seriously the ethics of human experimentation, and Institutional Review Boards (IRBs), organizations that must approve of research involving humans or animals, were created to help avoid unethical scientific experiments.

Why focus so much attention on a 60+ year-old study? On the one hand, it's important to recognize that there's a real sense in which Milgram's study takes place in virtual reality (in Antonin Artaud's sense of the term). Equally importantly, it's important to register that an experiment can become unethical because of how subjects respond to events in the study, like delivering electric shocks, that are themselves entirely simulated. Such experiments may also have important "after effects" on subjects that are hard to consent to in advance without introducing the possibility of demand characteristics (the risk of subjects acting to please, or frustrate, an experimenter when they think they know what an experiment is testing) (Nicholson 2011).[7] In Milgram's case, the learner's pain was simulated and scripted but because each subject believed that learners were really in pain, they responded to that suffering *as if* it were real and it was this fact that created ethical issues with the experiment protocol.

In this chapter I'll take a deeper look at experimental ethics especially when it comes to virtual reality.[8] The cost of VR hardware and the ease of developing VR simulations have changed radically over the last 20 years. Although it was only in 2016 that the first major commercial VR hmds became available (the HTC Vive and the Oculus Rift), moral psychologists have been exploring the potential of VR technologies long before this time. They have been seeking to expand not only the range of experiments they can do in the lab but also VR's potential to sidestep the thorny ethical problems that studies like Milgram's made so very obvious. AR is at a comparatively earlier stage of development and experiments are fewer and less complex in terms of simulating real-world decisions.[9] I begin by exploring the positive potential for VR in experiments. Because of their unique properties, VR and AR simulations can in fact lead us to make genuine improvements in the study of moral psychology because they allow experimenters to design experiments with much higher degrees of ecological validity than more traditional written experiments.

I'll then look at the ethical risks that these experiments can carry with them, focusing on the unique risks VR and AR introduce and that have been neglected by both experimenters and IRBs alike. I'll show that virtually real experiments pose nearly (and sometimes exactly) the same risks to human subjects as their real-life counterparts and that the ethics of using VR and AR simulations on human subjects needs to take this fact into account. This is a risk we ignore at our own peril.

I end the chapter by extending our look at the ethical features of non-experimental VR and AR applications, including game design and classroom use. If VR and AR are able to induce virtually real experiences in people then game designers would be wise to take into account perspectival fidelity and context-realism when they design simulations. Similarly, and especially in the wake of the COVID-19 pandemic, VR/AR simulations may help to close the gap between face to face and online education. For some courses, well designed educational simulations can help change how we learn. Educational VR/AR can, in specific contexts, not only function to mitigate the educational costs associated with the "emergency" virtual teaching that so many of us took part in, but can also radically expand access to education globally in transformative and powerful ways.

Virtual Reality and the Promise of Ecologically Valid Experiments

Moral psychologists study moral emotions, judgments, behavior, intuitions, and concepts. Moral psychology is thus a deeply interdisciplinary field. Philosophers, psychologists, neuroscientists, linguists, cognitive scientists, and sociologists all actively contribute to contemporary moral psychological research. Some researchers live in several worlds (e.g., experimental philosophers probe philosophical questions using the methods and tools of the social sciences) though most have, until recently, stuck to their own disciplinary tools and techniques to explore these concepts.

This short definition glosses over some of the most controversial issues in philosophy today: what makes something moral (as opposed to amoral) in the first place? Are there meaningful differences between moral intuitions and moral emotions? Are moral emotions the same as moral judgments? Must moral concepts necessarily include moral emotions as a part or are these two things independent? I won't be offering any answers to these questions here. Each of these questions has substantive and lively research programs devoted to them and disagreements run wide. The issues themselves are too deep and, I would argue, the science of moral psychology is too young for us to feel confident about any specific answers right now. Like any research, moral psychology has changed a lot in the 20th and 21st centuries both in terms of the tools and techniques used by researchers and in terms of the questions being asked by them.

For at least the last 40 years, a dominant tool used by moral psychologists has been the thought experiment. No doubt you're already familiar with some of these studies and these thought experiments – I've already discussed many of them in earlier chapters when discussing perspectival thought experiments. Trolley problems, ticking time-bombs, and dying violinists are commonplace. Such studies ask subjects what they think they would do (or should do) and a subject's response is supposed to help us learn more about what it is that's going on (psychologically, neurologically, morally) when we have to make moral judgments and even what moral judgments and moral concepts might be.

For the reasons already discussed in Chapter 2, the Thought Experiment Paradigm rests on several, likely false, assumptions about our abilities to imagine what we would (or what we think we *should*) do in perspectival thought experiments like these. Perspectival thought experiments assume that we're able to imagine both the conscious (doxastic) and non-conscious (subdoxastic) features of the situations they ask us to make decisions in and then imaginatively place ourselves in those situations to see what we would do in them. As we said back then, we're probably not able to do this (to successfully engage in "in-their-shoes" empathizing) if the situation that we're actually in is very different from the situation that we're being asked to imagine (i.e., if the situational features of these worlds are not what we called "base cases").

These problems help explain why it is that we're so bad at predicting our own future behavior when we're under moral pressure. For example, none of Stanley Milgram's test subjects believed that they were capable of giving lethal shocks to someone just because a man in a lab coat politely but firmly asked them to ... and yet in most of his studies, the majority of subjects did! This sort of imaginative perspective-taking failure is not uncommon. At the end of the second chapter, I hinted at the ability of VR and AR technologies to help us improve the Thought Experiment Paradigm by moving difficult background subdoxastic conditions out of the imagination and into the simulations themselves.

In this section I want to do two things. First, I want to argue that we really can (and should) seriously consider abandoning the Thought Experiment Paradigm as a way of discovering truths about what our naturalistic moral judgments (and moral concepts and moral emotions) look like. We should abandon it in favor of well-designed VR and AR simulations that faithfully recreate the situations called for within those thought experiments. Perspectivally faithful and context-real simulations are much more likely to show us what our real-world lived moral experiences are like than are any pen-and-paper thought experiments. Such simulations are also much more likely to have other advantages as well. For example, they're more likely to be better at containing what psychologists refer to as ecological validity.[10] Second, I'll pose an ethical dilemma for VR and AR moral psychological research: if we make our VR/AR experiments more ecologically valid then they're more likely to offer useful information, but they're also more likely to become unethical to carry out. If we continue working within the Thought Experiment Paradigm then, while we're not likely to run into ethical problems running our experiments, we're also not likely to gather ecologically valid data.

Experiments are always balancing acts between resources, ethics, and different kinds of experimental validity. While it's important that a researcher is confident that they've defined their terms well and that they have set up their experiment so that it measures what they hope it measures (i.e., that their experiments have good construct validity), it's equally important that the experiment itself is enough like the real world situations it's designed to study to leave us confident that the experimental results will actually help us predict real-world events and not just tell us about what subjects would do in the special context of a lab.

Imagine, for example, that we wanted to study juror behavior to help us better understand the factors that influence verdicts (something that, understandably, is actually the subject of intense research interest). How should we design such a study? One of the most common answers might be surprising. Like most experiments involving moral psychology, juror decisions are often studied by having subjects read about a case and then make a decision about guilt and sentencing (sentencing if subjects would find the defendant guilty), and they do this while sitting quietly in a lab.

Imagine that we were designing such a study. Imagine further that we're confident that we're accurately controlling for as many possible distractions (confounding variables) as possible. All subjects receive the same prompts and our lab spaces are identical. To avoid potential issues with priming, we might try to make sure that the experimenters explain the study in the same way to everyone and that we control for things like gender and racial bias. Let's also imagine that we could be reasonably certain that we're accurately recording each subject's responses (verdict, behavior, sentencing guidelines, etc.). Let's finally assume that we're scrupulous about the kinds of statistical analyses we apply to whatever data we collect (i.e., no p-hacking for us!) and that our results suggest our data is statistically significant relative to the standards of the best journals in our field. How confident should we be that we learned anything about how jurors actually behave? Are there any problems with a study like this?

We might be worried about the differences between the study environment and the environment that real-life juries work in. For example, real-life trials usually take at least hours (if not days ... or weeks) to finish and they require that jurors meet to talk among themselves about the case in a jury room (Breau & Brook 2007). Jurors, speaking from personal experience, also typically don't volunteer themselves to be on juries. Days are long. Meetings can be tense. Interpersonal conflicts can get in the way of dialogue and distractions are frequent. Even a bad lunch can affect our concentration or patience with others. All of these elements seem to be missing from thought-experiment style jury research.

There's at least some evidence to think that mock trials and real-life trials are different enough to make us doubt that we're learning much about real-world juror behavior from them (Wiener, Krauss, & Lieberman 2011). Our hypothetical study, while it might be rigorous in all the ways described, has a problem with ecological validity. Ecological validity is a way of talking about how well a study's environment is like the real-world environment that it's hoping to generalize its results to. While in some cases, ecological validity may not be very important to a study's results, in other cases it's especially significant. For an example of the former consider reflexive activity, like being startled. Reflexes might work the same in a lab as it does in your car or on your couch (Mauss & Robinson 2009) and so the fact that most reflexive activity doesn't take place in a lab or as a result of a researcher's prodding might not be relevant to the study of reflexive mechanisms (startles, for example, are surprisingly easy to generate even when subjects know they're about to be startled). However, moral judgments are probably not like startle reflexes. Ecological validity becomes especially important when we want to study moral judgment, moral behavior, and moral emotion.

The Thought Experiment Paradigm usually fares poorly when it comes to ecological validity. Most of our moral judgments aren't evoked by reading stories (short or long). The occasional email that contains an ethical dilemma aside, this isn't the context that most of our moral lives happen within. Instead, moral judgments usually take place in the context of being around other people, under time pressure, without perfect information, and under the influence of different (sometimes irrelevant) emotions. While thought experiments might not be good tools for simulating these conditions, VR and AR technologies seem designed to put people into (or augment) rich, immersive worlds. If designed well, these worlds might be far more ecologically valid, for experimental purposes, than what we're able to produce in our imagination when asked to think about thought experiments.

Virtually Real Experience in the Lab

Many of Stanley Milgram's subjects were harmed when they participated in his experiments. This is despite the fact that the shocks they thought they were delivering, and the learner's pain, were both virtual. Although full-scale real-life replications of Milgram's experiments are now, for ethical reasons, impossible to conduct in the lab (in most countries), psychologists eager to continue this research have hoped that transferring Milgram's experimental protocol from the real world to virtual worlds might make it ethical to replicate and re-examine his results. This assumption is worth challenging. Sometimes, virtual worlds can be ethically risky to place people too.

Psychologists like Mel Slater, for example, have claimed that VR technologies allow for "all social and psychological research where, for ethical or safety reasons, it is not possible to immerse experimental participants into the actual phenomena to be studied" (Slater et al. 2006). Slater is not alone in thinking this way. David Parsons, another psychologist, has claimed that "[t]he addition of virtual environments allows affective neuroscience researchers to move beyond the ethical concerns related to placing participants into real-world situations with hazardous contexts" (Parsons 2015). Psychologists are not alone in thinking along these lines. Neuroscientist Indrajeet Patil enthusiastically argued that

> [m]oral dilemmas are especially difficult to create realistically in laboratory settings because of the ethical problems associated with violent and harmful experimental situations. Virtual reality (VR) helps to take a step forward in studying such situations in a more ecologically valid manner.
>
> (Patil et al. 2014)

While VR and AR simulations can, as a result of their power to produce virtually real experiences, be a "step forward in studying such situations in a more ecologically valid manner," it's important for us to think through whether doing this reintroduces ethical issues into those simulations.

Fueled by the promise that VR can generate "presence" and that virtual experiences are not harmful, we are now in the beginning stages of a methodological VR renaissance in moral psychology (with AR not too far behind). However, this renaissance has already come with some costs. IRBs and many researchers currently operate under the belief that the simulated nature of VR and AR experiences doesn't raise ethically significant problems. No (real) harm, no (real) foul seems to be the operating assumption behind the researchers quoted above.

Institutional Review Boards, committees which must assess research for ethical risk before it can be conducted, have so far not taken seriously the idea that virtual experiences can be harmful. Given what I've said about perspectival fidelity, context-realism, and virtually real experiences, this is a mistake. Although VR and AR research is still at a relatively early stage, we already have evidence that virtually real experiences can be traumatic and that IRBs should be much more careful about how they think about the risks to human subjects in VR and AR research that has the potential to produce virtually real experiences.

Why think that VR and AR simulations can be harmful in the context of laboratory experiments? Mel Slater and his colleagues conducted an important early VR study that explicitly sought to replicate Milgram's experimental protocol, though translated into a VR CAVE environment (2006). Unlike the real-life replications of Milgram's protocol – which were not allowed to fully reproduce the study from beginning to end – because Slater's environment was "virtual" he was able to receive approval to replicate Milgram's protocols completely (including his use of a physical shock panel with switches going all the way to the "XXX Extreme Danger shock" level). The results of Slater's are, like Milgram's own results, fascinating:

> Some [subjects] giggled at the Learner's protests, as was observed by Milgram in the original experiments. When the Learner failed to answer at the 28th and 29th questions, one participant repeatedly called out to her "Hello? Hello? ..." in a concerned manner, then turned to the experimenter, and seemingly worried said: "She's not answering ..." In the debriefing interviews many said that they were surprised by their own responses, and all said that it had produced negative feelings ...
>
> (Slater et al. 2006)

Slater's subjects, by his own lights, seemed to behave a lot like Milgram's subjects did during the experiment. This was true not only in terms of their tendency to comply with the experimenter's request to continue giving the (virtual) learner more and more severe shocks but, as we can see above, subjects also mimicked the nervousness, anxiety, and trauma responses of Milgram's actual subjects.

Because Slater tried to faithfully replicate Milgram's experiments, he created a simulation with a high degree of perspectival fidelity and context-realism.[11] Using a CAVE VR system, Slater's subjects were led into a room where they sat in front of a real-life switchboard that they used to deliver their shocks. This setup faithfully replicates the haptics of Milgram's experiment as accurately as possible and in a situation that was context-real. Subjects could see the virtual learner sitting across from them projected on the wall within the CAVE. Although the virtual learner was not rendered as a high resolution character, she responded realistically to being shocked: she winced in pain, yelled and complained about the shocks using the same scripts written for Milgram's study, and, just like Milgram's learner, as the shocks became more intense she asked emphatically for subjects to stop the experiment. Slater noticed that

> [t]he participants in the [virtual version of the experiment] often behaved in a way that only made sense if they were responding to the virtual character as if she were real ... The voices of some participants showed increasing frustration at her wrong answers. At times when the Learner vigorously objected, many turned to the experimenter sitting nearby and asked what they should do.
>
> (Slater et al. 2006)

Slater even comes close to saying that at least some of his subjects had virtually real experiences while participating in his experiment. For example, Slater noticed that when the virtual learner stopped responding to their questions that his subjects started to do what most people do when dealing with a real-life person: they would repeat the question to see if the learner would respond to a second prompting. Slater asked himself:

> [w]hy did participants wait and repeat the question? This must be because this was not experienced as like watching a movie. Although individuals watching a horror movie may sometimes scream, or when watching a sports game on television may shout at the players, they do not expect that their actions can have any effect on the outcome of the movie or the game. Here, however, the situation was quite different. The actions of the participants actually mattered, and they behaved accordingly – they needed to wait, or withdraw altogether, in order to stave off or avoid the act of administering the shock and the unpleasant consequences that would follow from this.
>
> (Slater et al. 2006)

I think we have the resources to explain why Slater's subjects expected that their "actions actually mattered" and why they seemed to really believe, at least in the moment, that the virtual learner suffered each time they flipped the switch to shock her. Slater's VR experiment had all of the elements necessary to make it likely to generate virtually real experiences (at least for his most susceptible subjects). His observations help us see that his subjects actually had such experiences and that they were, predictably, "unpleasant." I'll return to this experiment later in the chapter. Another, even earlier, experiment on public speaking had a similar effect and is worth looking at for more evidence that virtually real experiences can be ethically significant.

In 2002 David Paul Pertaub conducted a VR study on public speaking. He wanted to figure out whether manipulating the reactions of a virtual audience (e.g., speaking in front of a hostile audience versus an engaged audience) would affect subjects. Pertaub asked his subjects to give a speech, giving them only a short time to prepare it, Pertaub described his study setup this way:

> A virtual seminar room was populated with an audience of eight male avatars seated in semi-circular fashion facing the speaker, all dressed in suits as if attending a formal meeting. These avatars were continuously animated, displaying random autonomous behaviors such as twitches, blinks, and nods that were consciously designed to foster the illusion of a real-life presence ... The room in which they were seated was a virtual counterpart of the real seminar room in which subjects completed their questionnaires.
>
> (Pertaub, Slater, & Baker 2002)

Notice that Pertaub seems to have designed his simulation to be as realistic as the technology of the time allowed. The virtual room was a copy of the room that subjects were in and his team made some effort to make sure the NPCs in the virtual seminar room reacted "as if attending a formal meeting" and acted in ways that mimicked the real life behaviors that subjects should expect of real audiences. The simulation, in other words, was high in context-realism and in perspectival fidelity. High in context-realism because the situation itself is fairly mundane (giving a speech in a seminar room) and audience reactions were intentionally programmed to mimic real-life behaviors (of bored or engaged audiences). The simulation was perspectivally faithful because it was not only in the first person but also did not include any other non-diegetic elements or graphical overlays.

Although Pertaub's study didn't involve the life-or-death consequences that Slater's study did, it nonetheless seemed to have a strong impact on his subjects:

> [S]ubjective reports confirmed that the negative audience was a strong anxiety-provoking experience and that it was capable of generating a range of very powerful emotions in the speaker.
>
> (Pertaub et al. 2002)

This should make sense to any of us who have been asked to give an unprepared speech in front of a hostile audience. The experience *really is* anxiety-provoking and the fact that Pertaub's subjects felt so much anxiety tells us that, at least in the moment, they must have been treating their experiences as if they were real.[12] Virtually real experiences aren't merely theoretical, they're being generated in virtual (VR and AR) environments in labs around the world now. However, because IRBs have yet to appreciate the risks to subjects of virtually real experiences, ethical safeguards to protect the very real human subjects involved in this research haven't yet been put in place. I'll circle back to this point at the end of this chapter.

More evidence that perspectivally faithful context-real simulations can induce virtually real experience comes from the burgeoning development of virtual reality exposure therapies (VRET). Traditional exposure therapies work by slowly acclimating a patient to an unpleasant stimulus (e.g., a phobic object or post-traumatic stress trigger) so that they can come to recognize early signs of panic, anxiety, and trauma and to develop cognitive strategies for managing these symptoms. Exposure therapies can be powerful tools for recovering from post-traumatic stress disorder, phobia, and other anxiety-based problems in living. However, in some countries (including the USA), access to such approaches can be prohibitively expensive without especially generous health insurance.

Virtual reality exposure therapies are designed to work on the same principles as traditional exposure therapy and are focused on treating the same kinds of problems in living (Rizzo et al. 2010; Rizzo et al. 2017). One of the most powerful pieces of evidence speaking to the strength of virtually real experiences comes from VRET studies. By their nature, VRET simulations are built to allow a patient to, slowly, increase a simulation's degrees of perspectival fidelity and context-realism to achieve therapeutic outcomes. They are, in that sense, a natural proof of concept for virtually real experiences.

An arachnophobe, for example, who enters a VRET simulation may find themselves seated in a context-real, perspectivally faithful therapist's office while they're exposed to very context-unreal stimuli (e.g., a cartoon spider). As they grow more comfortable and as they develop techniques for controlling their responses under these conditions, more context-realism can be introduced. Spiders can be made to appear more realistic (but at a greater distance) or their perspective can shift from the first to the third person. The goal, ultimately, is to introduce people to techniques designed to help them manage their responses to these triggers.

Simulations like these have shown great success. Importantly, VRET has demonstrated better therapeutic outcomes than imagined exposure therapies (where a client is asked to imagine being exposed to a trigger) and they have success rates closer to traditional exposure therapy (Opriş et al. 2012). Given the limitations of perspectival thought experiments, it's not surprising that imagined exposure therapies should have these shortcomings. As I argued in Chapter 2, we cannot consciously account for subdoxastic features of a situation without thereby robbing them of their subdoxastic effects. VRET studies are also powerful evidence that perspectival fidelity and context-realism really do help to modulate how people experience simulated worlds. The success of VRET strongly suggests that virtually real experiences are possible and that they can be controlled by varying a simulation's degree of context-realism and perspectival fidelity.[13]

What can we learn from these studies? VR really can help us get better, more accurate, data about what people really would do or how they would feel in the kinds of situations that, until recently, were dominated by the assumptions and methods built into the Thought Experiment Paradigm. Carefully designed VR and AR simulations high in context-realism and perspectival fidelity can give us a clearer picture of real-life moral psychology than thought experiments but this success comes with its own perils. We have to be as careful with the ethics of experiments that use VR or AR simulations as we would with experiments that put people into real-life moral dilemmas. Just as we, arguably, shouldn't be able to fully replicate Stanley Milgram's studies today, we shouldn't be able to put subjects into a context-real, perspectivally faithful simulation of Milgram's study without confronting many of the same ethical problems.

This doesn't mean that it's always (or even often) unethical to use VR and AR for research in moral psychology. One of the benefits of using these kinds of simulations is that experimenters have complete control over all elements of their design. As our understanding of the psychology of context-realism and perspectival fidelity increases researchers can become better at fine-tuning the degree of context-realism and/or perspectival fidelity of their simulations in order to avoid harming subjects while still being able to improve on the Thought Experiment Paradigm when it comes to data. What's needed is a better ethics for these AR and VR experiments.

In the next chapter I'll look at a few unique ethical problems that simulations can bring with them that extend beyond the risk of accidental trauma. For now, it's important for us to see that virtually real experiences are possible, that they go beyond just feeling present in a simulation (i.e., that they extend beyond location or action illusions), and that they can create ethically significant risks in an experimental context. As a philosopher and moral psychologist, the lab is a familiar place and the sudden widespread emergence in lab spaces of AR and VR simulations calls for immediate attention. IRBs should become more familiar with the risks of simulations on human subjects and act accordingly. There's another, very much larger space where VR has made sudden inroads: gaming. Any ethics of VR and AR needs to include non-lab uses of these technologies, including game environments.

Game Design

While researchers looking to make progress within moral psychology might want to maximize the degree to which their simulations produce virtually real experiences in their subjects (for the sake of collecting ecologically valid data), VR and AR game designers may prefer to do the opposite. If the point of a game is to be fun then a simulation that includes violence, murder, suffering, monsters, torture and the like probably won't be as fun as it could be (at least not to most consumers) if it triggers virtually real experiences in players. If war is hell, then a game whose purpose is to make hell fun shouldn't try to recreate it too faithfully.[14, 15]

Scott Stephan, director of games at FoxNext VR Studio (makers of *The Blu: Encounter* – an underwater VR simulation), recently offered some guidance to other game designers about good VR game design. For Stephan, VR games need to be developed differently from traditional "flat screen" games (games played on a two dimensional television or computer screen). Stephan cautions that his team has had to change how they think about developing certain kinds of experiences when they work in VR. For example, he argues

> that scary experiences, horror experiences need to be really finely calibrated. If you see a horror movie on a screen, you have the abstraction. It's not so frightening, and you know you're there for fun … I found that, in room-scale VR, things that might be fun on a TV screen, like jump scares …We actually have a rule that no creature should be larger than the size of a small dog. Anything above that and you get this primal, lizard-brain thing of, "Oh, this isn't a fun scare. It's a survival scare."
>
> (On immersive virtual reality 2019)

Earlier we noticed that Slater seemed to wonder why his subjects weren't treating the virtual learner in his experiment like they treat other fictional characters that they see on television or read in a book. Stephan here points to a similar kind of response that he and his beta testers had when they were learning how to design for VR (and by extension for AR as well): what works on a flat PC monitor may not work in an immersive VR world.

An experience that is bound to a screen is, by nature, less perspectivally faithful than an immersive experience of the very same content. Stephan supports my claim that these structurally unique features of VR and AR can really alter how we experience simulated worlds even if their content stays the same! A fun scare on a screen can easily become a virtually real survival scare in VR or AR, assuming that the simulation itself is context-real and perspectivally faithful (Luis' experience with hauntings in AR in the last chapter make for a good case study).

For these reasons, it's important that game designers incorporate knowledge of perspectival fidelity and context-realism into their workflows at every stage of development in a project. In part this benefits the artistic integrity of a gameworld: if the designer's intention is to have a "fun" scare moment, their intentions are thwarted if their world produces "survival" scares. As we'll see in a later chapter, adding elements to a simulation that increase its perspectival fidelity and context-realism but which aren't necessary for a simulation's narrative goals may not only thwart a game's internal narrative, it might traumatize its players, hurting sales and possibly even opening it to legal action.

Because user-psychology plays a role in virtually real experiences, it might also be necessary to add warnings aimed at especially sensitive users of the increased risks they would be exposed to if they were to play games especially high in perspectival fidelity and context-realism. At an upper limit, some games would be unethical (a perspectivally faithful and context-real torture simulator may thus be unethical even if a less faithful and less real version would not be).

I'll say more about the ethics of simulating evil in Chapter 6 as I apply a code of ethics for VR and AR simulation. Interested readers may want to head there before continuing. What I'll say there is that we can learn a lot about how not to design game worlds by paying as much attention to simulation structure (perspectival fidelity and context-realism) as we do to the actual content/story of a simulation's narrative. Keeping games fun means finding ways of engaging players without making them feel unintentionally alienated (games that might make players feel traumatized, sad, or angry).

Here it's enough to note that Stephan is right about horror experiences in VR and AR, but wrong to think that these concerns are limited to horror experiences. An unintentionally context-real, perspectivally faithful VR or AR game depicting genocide, rape, or torture is unlikely to find a large player base despite the fact that at least two of these themes appear frequently in PC games today. Game designers would be wise to keep those elements in mind throughout the development stages of a product. This is one of those times when something is both morally good and in our economic interest!

VR as Teaching Tool

As I write this, universities and secondary schools around the world, including my own, are struggling to cope with the impact of the COVID-19 pandemic. While many places have temporarily switched to online-only teaching out of a concern for public health, this option has proven unpopular with many students, parents, university administrators, and educators alike. It makes good sense to shift classes online given the risk of spreading the virus and the political realities that have shaped public health responses around the world. However, online education, though it has many benefits, also has important drawbacks and it's worth considering how we might shape a new future of online education using VR and AR technologies.

Until recently, the idea of using VR or AR in education was relatively rare. Not only were these technologies economically beyond the reach of most families and school systems, but designing for the VR or AR systems that existed at the time was slow and expensive. Although commercial VR systems have drastically reduced the cost of entry into the VR marketplace, today's commercial VR and AR headsets are still beyond the reach of most consumers. The cost of a VR headset *and* a VR-capable computer can easily eat more than half of the average American household's monthly income (United States Census Bureau 2017) and the situation is worse in less well-off places. We are not quite at the level of cheap and ubiquitous VR or AR.

Despite this, things are looking promising for the future. In 2019 Mark Zuckerberg announced that Facebook's VR company, the Oculus Corporation, had sold over $100,000,000 in content sales during the fiscal year (Matney 2019). The same year also saw VR's first "killer app" (a game called *Beat Saber*) sell over 1 million copies (Roettgers 2019). Oculus is only one of several major players in the VR market and its growth is not unusual. Even more impressively, Niantic's AR game, *Pokemon Go!*, has, since its release in 2016, generated nearly $4 billion in sales (Desatoff 2020).

With the surge of commercial VR and AR successes, it's more and more likely that the cost of VR and AR systems will keep dropping. Nearly half of the world's population now owns a smartphone (Number of smartphone users 2019) so it's not too big a leap to imagine hundreds of millions of VR and AR users entering the marketplace over the next two decades as the price of these technologies drops to comparable levels.

Beyond concerns about making emergency online pandemic teaching better, does it make sense to take seriously the idea that VR and AR can actually help us *improve* education and training? I think we have some evidence to think that the answer is yes. Preliminary research has suggested that VR and AR can be useful for:

- Controlling attention in students with ADHD (Parsons et al. 2007)
- Increasing motivation in science education (Liou & Chang 2018)
- Improving perception for spatial skills learning (Kim & Irizarry 2020)
- Enhancing education using novel forms of interactivity (Sharma, Agada, & Ruffin 2013)
- Improved approaches to language learning (Parmaxi & Demetriou 2019)
- Using virtual tourism to allow students to better connect their classroom religious studies with religious sites (Johnson 2018)
- Increased motivation for learning (Erbas & Demirer 2018)[16]
- Corporate compliance training (Schmid Mast et al. 2018)
- Surgical training (Thomsen et al. 2017)

Unlike the moral psychology lab, where our goal is to collect ecologically valid realistic responses, our goal with educational VR and AR doesn't always rely on users having virtually real experiences. Educational VR and AR have a much larger design space, more design options, to meet their goals. Recall Samaya's drafting program in the last chapter. Samaya's program can be a good educational program without depending on her thinking and feeling as if she's really hovering near an unfinished bridge. In fact, those feelings might get in the way of it *being* a successful program. A program like Samaya's can easily become a teaching tool for new civil engineers to design and test designs as they learn core concepts about structural engineering.

Similarly, AR teaching tools can not only help bring additional teaching materials to the physical classroom environment (flash cards, pop-up directions, educational mini-games), they can also help in other ways. For example, we might imagine a language learning lab that can produce, via AR, real-time transcripts of conversations to help students improve their own reading, writing, and speaking skills. AR-augmented chemistry labs can help students better visualize and understand chemical reactions and compounds via interactive three dimensional representations.

The same goes for VR in the classroom. I first started using VR when I was teaching a class on "Philosophical issues in virtual reality" in the spring of 2019. In that class we spent about a third of our time in a VR lab getting first-hand experience of how VR can be used for moral psychology, gaming, virtual tourism, historical re-enactment, exposure therapy, and so on. Students very quickly lost themselves in virtual worlds and often forgot that they were still very much surrounded by their peers. During a week when we were exploring the possibilities of VRET for treating phobias, one student even (unintentionally!) swatted me in the face to shoo away a virtual spider. I've since expanded how I use VR to include bringing it into the classroom when I teach specific subjects in classes on ethics, neuroethics, and even classes on emotion. In an unpublished pilot study on the effects of using VR to teach philosophical concepts I've begun work on, students in VR enhanced classrooms have, thus far, tended to more enthusiastically approach course content and tend to remember their experiences for a longer period after our term is over.[17]

As VR and AR become easier and cheaper to use and incorporate into teaching spaces, it can become even easier to get more out of both physical and online education. One of the major benefits of online learning is that it's widely accessible and scalable. Universities are global institutions, and a student living in a country like Taiwan can join a virtual classroom in California as easily as a native Californian (though they'll still have to wrestle with the time difference if the class meets synchronously).

Depending on our educational aims, we may or may not want or need to create virtually real simulations. For example, skill-based knowledge (dance, lab-based classes, and so on) would benefit from at least some degree of virtual realism in order to teach precise movements, reactions, or skills. Surgical training, for example, often aims to realistically model not only the context of the operating room but also realistic haptics during surgery and has demonstrated a degree of success with improving the surgical skills of novices (Thomsen et al. 2017). Such simulations have the added benefit of allowing users to review their own work in context-unreal ways (floating cameras, close-ups from the point of view of the scalpel, etc.) that can augment their education by better pinpointing places to improve. We can imagine simulations aiming to teach form and position for dance or those aiming to teach laboratory or field research skills being modeled on programs like the surgical trainer. Because they're infinitely iterable (once you have the simulation built it can be shared with anyone who has access to VR or AR technology), these teaching tools scale well.

Content-based knowledge (e.g., knowing "that" instead of knowing "how") need not require the design of simulations capable of producing virtually real experiences at all so long as someone learns the relevant pieces of knowledge and can transfer that knowledge to new situations. Allowing students to fly through a VR simulation of the Buddhas of Bamiyan (destroyed by Taliban forces in 2001) can help them better engage with and then better recall and understand the tenets of early Buddhism in the Middle East (Johnson 2018) while a simulation of Socrates' trial and (spoiler!) conviction may help students better appreciate, remember, and apply what they learned about Greek virtue ethics. In these cases, because we're not interested in seeing how students would *really* have acted in Athens, perspectival fidelity and context-realism only matter to the degree to which they get in the way of our educational goals. In a later chapter I'll consider additional ethical problems that might be a part of what are called VR or AR "educative" nudges, but those are unlikely to be a problem for us here (educative nudges are often invisible to users in a way that's not usually a part of traditional education).

Should we be worried about VR and AR as teaching tools? Virtually real experiences can definitely get in the way of education, and educators (and software engineers) should be cautious about how they design these simulations. A virtually real experience of the trolley problem is likely to be too traumatic and unethical to use in the classroom (as would be virtually real simulations of atrocities in all but the most specific and constrained uses). VR simulations can also, like all media we use in the classroom, reflect racist, ableist, misogynist, and classist norms based on what they're about and whose point of view they represent (or leave out). In that sense, the traditional care that good educators already ought to bring to their classrooms must also be brought to the virtual worlds they ask their students to enter.

The Equivalence Principle

It seems likely that many users who enter a simulation high in context-realism and perspectival fidelity will have virtually real experiences. Virtually real experiences can be really helpful. In the context of sessions of virtual reality exposure therapy, they can help a person better control their post-traumatic stress triggers or learn techniques for mitigating their phobias. In the context of education, they can help us more quickly learn skills (especially compared to online labs that use videos or texts in place of hands-on learning).

However, in other contexts, virtually real experiences can hurt us, and members of IRBs, educators, VR and AR designers, and commercial users, owe it to themselves to avoid causing unnecessary harm to others. For several years, I have been advocating that Institutional Review Boards and software designers adopt an ethical principle called The Equivalence Principle (Ramirez 2018; Ramirez & LaBarge 2018; Ramirez 2020; Ramirez & LaBarge 2020).

> The Equivalence Principle (TEP): If it would be wrong to subject a person to an experience, then it would be wrong to subject a person to a virtually real analog of that experience. As a simulation's likelihood of inducing virtually real experiences in its users increases, so too should the justification for the creation and use of the simulation.

At heart TEP is a straightforward moral principle. If we believe that we should treat similar harms similarly then what's needed is the argument that virtually real harms affect a person in similar ways as real-life harms do. Given the experiences that Pertaub's subjects had while giving their public speeches, given the anxiety and stress exhibited by Slater's subjects as they shocked a virtual learner, and given the successes of VRET, I think it's not too early to draw at least preliminary conclusions about the power of virtually real experiences.

An Institutional Review Board that uses TEP and that has an understanding of perspectival fidelity and context-realism would not have approved of Slater's replication of Milgram's obedience studies. Since Milgram's initial experiments, IRBs have been very strict about how replications of this study can be carried out. They have, rightly, decided that complete replications of Milgram's protocol would be unethical to carry out, knowing what we know now about the stress and trauma that the studies imposed on their subjects. Suppose, however, that someone thought TEP couldn't possibly be right. In Milgram's studies, even though nobody was *really* being harmed, subjects believed that they were hurting a real-life person and the first-hand accounts of many of these subjects support this interpretation. A critic might say that, in Slater's replication, it's impossible to say the same thing. An argument like this may very well have been made when the IRB responsible for approving Slater's replication gave it the green light. I think there are two responses that we can make to this concern.

First, I think it's too early for us to dismiss the claim that people believed they were hurting someone in Slater's study. As Slater himself notes, his subjects really seemed to feel and act *as if* their experiences were really happening. We might get some help here by turning to a distinction introduced by philosopher Tamar Szabó Gendler between beliefs and aliefs (2008).[18]

To the philosophically uninitiated, this might seem a strange distinction. While *beliefs* are conscious, fact-asserting mental states (e.g., we believe *that*), *aliefs* are subdoxastic (non-conscious) emotional and behavioral associations that can be said to represent something in the world. Yelling in fear during a horror movie implies that the viewer has the *alief* that they're in danger even though she almost certainly doesn't *believe* that she's in actual danger at all. In this case the alief is a behavioral association between emotion, vocalized motor activity (e.g., yelling), and perception. In a case like this, where someone seems afraid of something they know can't really hurt them, Gendler argues that

> it seems clear that the subject believes the walkway is safe, that the substance is edible or potable, that the wallet is in New Haven, that the theater is in no danger of being invaded by slime, and so on. Ask the subject directly and she will show no hesitation in endorsing such claims as true. Ask her to bet, and this is where she will place her money. Ask her to think about what her other beliefs imply and this is what she will conclude. Look at her over arching behavior and this is what it will point to. At the same time, the belief fails to be accompanied by certain belief-appropriate behaviors and attitudes: something is awry.
>
> (Gendler 2008, 637–638)

One way of responding to the critique of TEP is to defend virtually real experiences by showing that perspectively faithful and context-real simulations are especially powerful alief generators when compared to other kinds of simulated experiences (movies, books, the imagination, etc.). Slater, recall, noted that his subjects weren't treating their simulated experiences like they normally treat theatrical or novelistic simulations. Aliefs can help us explain why.

Someone who alieves that she is murdering another person suffers a similar trauma to that of someone who believes that they've really murdered someone (i.e., in this case someone who both alieves and believes that they've murdered someone). Notice that Milgram's study is structurally similar to this case. Milgram's study caused trauma as a result of the aliefs that his subjects formed during the study (we can see this because in many cases the trauma remained even after some subjects were told that nobody was really hurt and that the learner was actually in on the study all along).

However, I think it's too early to concede the point about aliefs and beliefs. It may very well be that *in the moment* when Slater's subjects were part of the study that they not only *alieved* that they were murdering the learner but they *believed* it too. This is probably the best way for us to make sense of the subject who, according to Slater, looked at him "in a concerned manner, then turned to the experimenter, and seemingly worried said: 'She's not answering'" (Slater et al. 2006, 5). This subject, at least, is not merely relying on subdoxastic aliefs to suggest virtually real concern for the learner. They seem to genuinely *believe* that something is wrong and that someone might be seriously hurt. It seems possible that this subject may very well have had the right mix of psychological dispositions that made their experience more virtually real than most and hence triggered the formation of beliefs and aliefs. TEP, if used judiciously by IRBs approving VR and AR research, would caution against including such a subject in a study like this.

But suppose our critic isn't convinced. I think we have a second reply to concerns about TEP and studies like Slater's replication of Milgram. Milgram's original experimental subjects were not only harmed by the stresses of the experiment and by believing (and/or alieving), even if only during the experiment, that they were hurting someone. While that's an important part of why Milgram's studies are unethical to faithfully replicate today, it isn't the whole story. Beyond the immediate trauma, Milgram's subjects were also harmed by being burdened with a specific kind of knowledge: that they're the kind of person who would give in to authority even to the point of murder. Notice that this piece of information is true about the subject even though they didn't actually hurt anyone, even though it was all play-acting on the part of the learner. For example, one of Milgram's subjects seemed more troubled about what they learned about themselves than the subjective trauma they experienced during the study:

> It has bothered me that I went all the way ... I am a person who has had the opportunity to do a great deal of reading and reflecting ... I form my own opinions from the newspapers and mass periodicals ... I also read the smaller circulation publications such as The Reporter, The Nation, The New Republic, and similar publications. I consider myself better informed and I hope more cultured than the average non-college student. In spite of all this I gave the same performance that the average slob, taken off the street, would probably have done. This I consider frightening.
>
> (Reaction of subjects, 1962, quoted in Nicholson 2011)

The same conditions hold for Slater's (2006) VR replication of Milgram's experiments. Once the dust is settled, the experiment is over, and subjects are debriefed, it's still possible that their sense of who they are might be permanently altered. While the data gathered by Milgram and Slater might be socially beneficial, IRBs are usually loath to approve research where all of the risks are borne by the experimental subjects without sharing in the benefits.[19, 20]

The Code of Federal Regulations contains guidelines intended to help IRBs make decisions about how to approve research that involves human subjects. Guideline 45 CFR 46 §46.111 specifically advocates that in order to approve research covered by this policy the IRB shall determine that all of the following requirements are satisfied:

1   Risks to subjects are minimized:

    i   By using procedures that are consistent with sound research design and that do not unnecessarily expose subjects to risk, and

2 Risks to subjects are reasonable in relation to anticipated benefits, if any, to subjects, and the importance of the knowledge that may reasonably be expected to result. In evaluating risks and benefits, the IRB should consider only those risks and benefits that may result from the research (as distinguished from risks and benefits of therapies subjects would receive even if not participating in the research). The IRB should not consider possible long-range effects of applying knowledge gained in the research (*e.g.*, the possible effects of the research on public policy) as among those research risks that fall within the purview of its responsibility.

(45 CFR 46 §46.111 2009)

If I'm right about the effects of virtually real experiences, then simulations that can generate them will introduce ethically significant risks that experimenters should minimize in any study involving human subjects. IRBs should incorporate TEP, or a principle like it, to help them better assess the risks that VR studies will pose. Researchers can control, to at least some degree, all three elements that combine to create virtually real experiences. Because they are entirely responsible for a simulation's design, they can work to minimize risk by varying their simulation's degrees of perspectival fidelity and context-realism to the bare minimum required to collect the data they're interested in. Because they can screen subjects before running them through their study, researchers can also make sure that people most susceptible to virtually real experiences are carefully monitored (or, if the risks are high enough, excluded from the study altogether).

Game designers can also use TEP to help avoid unintentionally traumatizing their players. In Chapter 6 I'll use a controversial mission from gaming company Rockstar North's latest (2013) version of their *Grand Theft Auto* franchise as a case study. It turns out, I'll argue, that some games (even traditional PC-based games like *Grand Theft Auto*) can unintentionally be created so that they're high enough in perspectival fidelity and context-realism to negatively affect their players. Because it's almost always in a gaming company's interest to keep their player base coming back for more, TEP can be a helpful tool for avoiding accidental trauma (and the moral outrage it can generate) in players. For example, intentionally adding elements that diminish perspectival fidelity (non-diegetic soundtracks, graphical overlays, third-person perspective) and context-realism (unrealistic physics, superhuman stamina, less reasons-responsive NPCs) can help players enjoy content that would be otherwise experienced as immoral or traumatic.

Conclusion

Virtually real experiences are powerful. It's because of the ability of AR and VR technologies to give their users virtually real experiences that gives them the potential to overturn the Thought Experiment Paradigm in moral psychology and to help usher in a new era of ecologically valid research on moral judgment, moral emotion, and moral concepts. However, because virtually real experiences are treated *as if* they were real (at the very least in the alief sense if not the belief sense), they introduce a host of new ethical issues to any experiment that uses VR and AR simulations to study human behavior.

Simulations high in perspectival fidelity and context-realism deserve special attention from IRBs when it comes to assessing the risk of harm that they pose to potential subjects, especially subjects with the right mix of psychological dispositions. In many cases, the risk of harm in a simulated world can be the same (or very nearly the same) as a real-world experiment. Studies like those by Mel Slater and David Pertaub and the success of virtual reality exposure therapies suggest that the ethics of virtual experiences needs to take virtually real experiences more seriously. The Equivalence Principle, which suggests that the ethical justification for exposing someone to a simulated experience should vary in proportion to that simulation's ability to generate virtually real experiences, is my attempt to fill in the gap that exists in experimental ethics. Although some psychologists seem to think that virtual experiences can't become ethically significant, TEP suggests when (and why) this isn't true.

Game designers also need to become more aware of how the simulated worlds they build can work to accidentally traumatize their players (and thus how they can unintentionally design a game that's less engaging and less fun than it could be). A better understanding of perspectival fidelity, context-realism, and TEP can go a long way toward making games that are not only more ethical but more fun.

I've focused this chapter on the trauma that virtually real experiences can cause. Although important, VR-induced trauma is only one of the ethical risks accompanying simulation use and design. In the chapters that follow, I'll look at two different possible harms associated with VR and AR simulations. In Chapter 5 I'll focus on manipulation and deception. Simulations that promise to show us what it's like to be someone else, even when well intentioned, deceive us and manipulate us. This becomes especially problematic when they're used to correct for social injustices stemming from intersectional oppression. I'll argue in that chapter that simulations like those can often enact intersectional injustice in the name of alleviating it. In Chapter 6 I'll condense these insights in order to provide a code of ethics for VR and AR design. I'll not only incorporate the codes of ethics produced by professional organizations like the IEEE and the ACM, but I'll extend their analyses to offer specific design considerations unique to VR and AR technologies.

Notes

1 Milgram's experiments are not without their defenders. For more on how, and why, Milgram's experiments were initially criticized and the back-and-forth discussions of the experiments' ethics, see Nicholson (2011).

2 In that sense, Milgram's experiments remain valuable today as we confront a 21st century with active genocides and instances of ethnic cleansing taking place around the globe to decidedly low outrage.

3 In other words, Milgram's subjects were led to believe that they were randomly selected to play the part of the teacher and that they could very well have been chosen to play the part of the learner. As mentioned above, this was a lie. All learners were actors and selection of roles was not random.

4 There is, of course, some harm that might be ethical to include as a part of a study. *The Belmont Report* (NCPHS 1978) and the US Federal Code of Regulations (Protection of human subjects 2009) along with the American Psychological Association's own ethical guidelines, all attempt to spell out when a risk or harm to a subject in an experiment is ethically justifiable. What they all agree on is that Miglram's studies did not meet current standards.

5 Quoted in Nicholson 2011.

6 There are a few partial replications out there that attempt to remove the most ethically egregious elements of Milgram's protocols so that they might get approval from institutional review boards but these are not, strictly speaking, replications of his experiment. Most studies replicate only a part of Milgram's protocols (Burger 2009; Doliński et al. 2017). In both cases, changing this variable changes the study in important ways. Importantly, Milgram didn't appear to debrief all of his subjects. One experimental subject said "I felt so bad afterward." Another remarked "I wanted to call and actually apologize … and I went to the telephone book and looked up the name – he used the name Richardson or something and unfortunately there were three of them with exactly the same first and last name [so] I didn't make the call" ("Subjects' conversation" 1963). This further augments the ethical problems with Milgram's study, though even *with* a debriefing session, the experiment remains unethical.

7 These aspects of an experiment connect with the literature on so-called "transformative experiences" (Paul 2014). Transformative experiences are experiences, like becoming a parent, whose effects are said to be impossible to understand prior to direct experience and which can complicate the traditional picture of experimental ethics, medical consent, and decision theory (Pettigrew 2015).

8 AR experiments are fewer and farther between, but as mentioned in the prior chapter, because AR simulations can get many elements of perspectival fidelity "on the cheap" they not only are capable of raising the same sets of ethical issues but they're more likely to do so without additional care in simulation design.

9 For example, Ratna Sari and colleagues (2021) conducted a simplified experiment using AR moral dilemmas to investigate whether these simulations improved a construct they call "moral imagination" (the ability to identify relevant moral variables and stakeholders). These simulations are, however, very low in context-realism and perspective fidelity, and so it's not clear that studies like this one are generating virtually real experiences (though for the same reasons, it's unlikely that the data gathered is ecologically valid).

10 I'll say more about this in a moment. Ecological validity is only one of several forms of validity that an experiment can be said to have. To design an ecologically valid experiment is to have an experimental situation that contains the same sorts of features that the real-world context it aims to model also contains. Not all experiments actually need to be ecologically valid; however, experiments in moral psychology probably do.

11 We don't know much about the subjects in Slater's study, especially in terms of their psychological dispositions (Big Five traits) and their degree of dissociation. Such factors may explain why some subjects had more powerful virtually real experiences than others throughout the course of the study (though this is conjecture on my part).

12 It's important to keep in mind that virtually real experiences are those experiences that subjects treat as if they were real in the moment and not necessarily after the experience is over. Additionally, as with the subjects in Slater et al.'s (2006) study, we don't have any information about Pertaub's subjects and their psychological predispositions or personality traits. Here too it may be that subjects who reacted most strongly did so as a result of the interactions between the simulation's features (perspectival fidelity, context-realism) and their own psychological dispositions.

13 Augmented reality exposure therapy (ARET) is, as with most things AR, at an earlier stage of development though it too shows promising results as an approach to treating specific phobia (De Witte et al. 2020). Interestingly, De Witte and colleagues found that subject responses were strongest depending on the "perceived realism" of the phobic object (virtual spiders in this case) and hence are in line with VRET studies and my account of virtually real experience.

14 I'll leave aside, for now, the question about the ethics of simulated evil. Maybe games, especially VR games, *shouldn't* try to make war fun. I'll focus on this question in a later chapter when I discuss what's called "The Gamer's Dilemma" (Luck 2009).

15 By the same token, a military training simulation may aim to do the exact opposite! By more carefully recreating the real-life conditions faced by military personnel, VR and AR training may help improve not only a soldier or officer's performance but may also be useful in terms of identifying who (and when) to deliver psychological services to before a soldier experiences long-lasting trauma. We're now at the point where such questions are not merely hypothetical. In 2021, the Microsoft Corporation was awarded a contract worth up to $22 billion to develop its Hololens for use with the US Army (Novet 2021).

16 Admittedly, the authors did not find a difference in learning outcomes in this particular study despite the increase in motivation. Further work should be done to explore whether the increase in motivation was due to the novelty of AR (novelty effects) or whether the AR interventions themselves increased motivation.

17 This could also be an effect of sheer novelty. VR and AR are new and exciting intrusions into traditional learning environments and it might be the newness that's sticking in their minds. More and better data needs to be collected and analyzed. I include my pilot study anecdote here to encourage others who have access to VR equipment to experiment with it too. My research team and I have created several free-to-use moral dilemmas including Trolley Problems, Violinist Dilemmas, and Clinical Ethics simulations. Anyone with an internet connection and VR equipment can download them at https://www.erickjramirez.com/virtual-reality.html

18 Whether or not aliefs are a distinct psychological kind is, unsurprisingly, a matter of intense debate (Albahari 2014).

19 The Belmont Report, one of the hallmark achievements in American bioethics, explicitly cautions both researchers and IRBs that "[t]he method of ascertaining risks [to subjects] should be explicit, especially where there is no alternative to the use of such vague categories as small or slight risk. It should also be determined whether an investigator's estimates of the probability of harm or benefits are reasonable, as judged by known facts or other available studies" (NCPHS 1978). I argue that researchers and IRBs are not doing a good job of accurately assessing the risks of perspectivally faithful and context-real simulations.

20 Chapter 6 will continue to focus on the issue of virtual experience, self knowledge, and behavior change, to help us make sense of recent issues in simulation and experimental ethics.

References

45 CFR 46 §46.111 (2009).

Albahari, M. 2014. Alief or belief? A contextual approach to belief ascription. *Philosophical Studies*, 167, 701–720.

Breau, D.L., & Brook, B. 2007. "Mock" mock juries: A field experiment on the ecological validity of jury simulations. *Law and Psychology Review*, 31, 77–92.

Burger, J. 2009. Replicating Milgram: Would people still obey today? *American Psychologist*, 64 (1), 1–11.

Desatoff, S. 2020, July 6. Pokemone Go surpasses $3.6 billion in lifetime revenue. GameDaily. Retrieved from https://gamedaily.biz/article/1795/pokemon-go-surpasses-36-billion-in-lifetime-revenue

De Witte, N.A.J., Scheveneels, S., Sels, R., Debard, G., Hermans, D., & Van Daele, T. 2020. Augmenting exposure therapy: Mobile augmented reality for specific phobia. *Frontiers in Virtual Reality*, 1, 8.

Doliński, D., Grzyb, T., Folwarczny, M., Grzybala, P., Krzyszycha, K., Martynowska, K., & Trojanowski, J. 2017. Would you deliver an electric shock in 2015? Obedience in the experimental paradigm developed by Stanley Milgram in the 50 years following the original studies. *Social Psychological and Personality Science*, 8 (8), 927–933.

Erbas, C., & Demirer, V. 2018. The effects of augmented reality on students' academic achievement and motivation in a biology course. *Journal of Computer Assisted Learning*, 35, 450–458.

Gendler, T.S. 2008. Alief and belief. *Journal of Philosophy*, 10, 634–663.

Hollander, M., & Maynard, D.W. 2016. Do unto others … ? Methodological advance and self- versus other-attentive resistance in Milgram's "obedience" experiments. *Social Psychology Quarterly*, 79 (4), 355–375.

Johnson, C.D.L. 2018. Using virtual reality and 360-degree video in the religious studies classroom: An experiment. *Teaching Theology and Religion*, 21, 228–241.

Kim, J., & Irizarry, J. 2020. Evaluating the use of augmented reality technology to improve construction management students' spatial skills. *International Journal of Construction Education and Research*, 17 (2), 99–116.

Liou, W., & Chang, C. 2018. *Virtual reality classroom applied to science education.* 23rd International Scientific-Professional Conference on Information Technology (IT), Zabljak, Montenegro. pp. 1–4. doi:10.1109/SPIT.2018.8350861.

Luck, M. 2009. The gamer's dilemma: An analysis of the arguments for the moral distinction between virtual murder and virtual paedophilia. *Ethics and Information Technology*, 11, 31–36.

Matney, L. 2019, September 25. Oculus eclipses $100 million in VR content sales. *Tech Crunch.* Retrieved from https://techcrunch.com/2019/09/25/oculus-eclipses-100-million-in-vr-content-sales/

Mauss, I.B., & Robinson, M.D. 2009. Measures of emotion: A review. *Cognition and Emotion*, 23 (2), 209–237.

NCPHS (National Commission for the Protection of Human Subjects of Biomedical and Behavioral Research). 1978. *The Belmont Report: Ethical principles and guidelines for the protection of human subjects of research.* Bethesda MD: NCPHS.

Nicholson, I. 2011. "Torture at Yale": Experimental subjects, laboratory torment and the "rehabilitation" of Milgram's "Obedience to Authority." *Theory and Psychology*, 21 (6), 1–25.

Novet, J. 2021. March 31. Microsoft wins U.S. Army contract for augmented reality headsets, worth up to $21.9 billion over 10 years. CNBC. Retrieved from https:// www.cnbc.com/2021/03/31/microsoft-wins-contract-to-make-modified-holo lens-for-us-army.html

Number of smartphone users. 2019, September. https://www.statista.com/statistics/ 330695/number-of-smartphone-users-worldwide/

On immersive virtual reality technology and the ethics questions it raises. 2019, October 8. *VR Life*. Retrieved from https://www.vrlife.news/immersive-experien ces-virtual-reality-design-ethics/

Opriş, D., Pintea, S., Garcia-Palacios, A., Botella, C., Szamosközi, Ş., & David, D. 2012. Virtual reality exposure therapy in anxiety disorders: A quantitative meta-analysis. *Depression and Anxiety*, 29 (2), 85–93.

Parmaxi, A., & Demetriou, A. 2019. Augmented reality in language learning: A state-of-the-art review of 2014–2019. *Journal of Computer Assisted Learning*, 36, 861–875.

Parsons, T.D. 2015. Virtual reality for enhanced ecological validity and experimental control in the clinical, affective and social neurosciences. *Frontiers in Human Neuroscience*, 9 (650). DOI: 10.3389/fnhum.2015.00660.

Parsons, T.D., Bowerly, T., Buckwalter, J.G., & Rizzo, A.A. 2007. A controlled clinical comparison of attention performance in children with ADHD in a virtual reality classroom compared with standard neuropsychological methods. *Child Neuropsychology*, 13, 363–381.

Patil, I., Cogoni, C., Zangrando, N., Chittaro, L., & Silani, G. 2014. Affective basis of judgment–behavior discrepancy in virtual experiences of moral dilemmas. *Social Neuroscience*, 9 (1), 94–107.

Paul, L. 2014. *Transformative experience*. Oxford: Oxford University Press.

Pertaub, D.P., Slater, M., & Barker, C. 2002. An experiment on public speaking anxiety in response to three different types of virtual audience. *Presence Tele-operators and Virtual Environments*, 11 (1), 68–78.

Pettigrew, R. 2015. Transformative experience and decision theory. *Philosophy and Phenomenal Research*, 91 (3), 766–774.

Ramirez, E. 2018. Ecological and ethical issues in virtual reality research: A call for increased scrutiny. *Philosophical Psychology*, 32 (2), 211–233.

Ramirez, E. 2020. How to (dis)solve the Gamer's Dilemma. *Ethical Theory and Moral Practice*, 23, 141–161.

Ramirez, E., & LaBarge, S. 2018. Real moral problems in the use of virtual reality. *Ethics and Information Technology*. https://doi.org/10.1007/s10676-018-9473-5

Ramirez, E., & LaBarge, S. 2020. Ethical issues with simulating the Bridge problem in VR. *Science and Engineering Ethics*, doi:10.1007/s11948-020-00267-5.

Rizzo, A., Difede, J., Rothbaum, B.O., Reger, G., Spitalnick, J., Cukor, J., & Mclay, R. 2010. Development and early evaluation of the Virtual Iraq/Afghanistan expo-sure therapy system for combat-related PTSD. *Annals of the New York Academy of Sciences*, 1208, 114–125.

Rizzo, A., Roy, M.J., Hartholt, A., Costanzo, M., Beth Highland. K., Jovanovic, T., Norrholm, S.D., Reist, C., Rothbaum, B., & Difede, J. 2017. Virtual reality applications for the assessment and treatment of PTSD. In S.V. Bowles & P.T. Bartone (Eds.), *Handbook of Military Psychology*. Cham, Switzerland: Springer International, 453–471.

Roettgers, J. 2019, March 14. Beat Saber sells more than one million copies, releases first music pack. *Variety*. Retrieved from https://variety.com/2019/digital/news/beat-saber-one-million-copies-sold-1203163174/

Sari, R.C., Sholihin, M., Yuniarti, N., Purnama, I.A., & Dwi Hermawan, H. 2021. Does behavior simulation based on augmented reality improve moral imagination? *Education and Information Technologies*, 26, 441–463.

Schmid Mast, M., Kleinlogel, E.P., Tur, B., & Bachmann, M. 2018. The future of interpersonal skills development: Immersive virtual reality training with virtual humans. *Human Resource Development Quarterly*, 29, 125–141. https://doi.org/10.1002/hrdq.21307

Sharma, S., Agada, R., & Ruffin, J. 2013. *Virtual reality classroom as a constructivist approach*. IEEE Southeastcon, Jacksonville FL. pp. 1–5. doi:10.1109/SECON.2013.6567441.

Slater, M., Antley, A., Davison, D., Swapp, D., Guger, C., Barker, C., Pistrang, N., & Sanchez-Vives, M.V. 2006. A virtual reprise of the Stanley Milgram obedience experiments. *PLoS ONE*, 1, e39.

"Subjects' conversation" (1963, February 28). Stanley Milgram Papers (Series II, Box #44). Yale University Archives, New Haven CT.

Thomsen, A.S.S., Bach-Holm, D., Kjærbo, H., Højgaard-Olsen, K., Subhi, Y., Saleh, G.S., Park, Y.S., la Cour, M., & Konge, L. 2017. Operating room performance improves after proficiency-based virtual reality cataract surgery training. *Ophthalmology*, 124 (4), 524–531. https://doi.org/10.1016/j.ophtha.2016.11.015

United States Census Bureau. 2017, September 14. Median household income in the United States. Retrieved from https://www.census.gov/library/visualizations/2017/comm/income-map.html

Wiener, R., Krauss, D.A., & Lieberman, J.D. 2011. Mock jury research: Where do we go from here? *Behavioral Science and the Law*, 29, 467–479.

5   Why It's Unethical to Use VR and AR as "Empathy" Machines

Some years ago, a group of well-intentioned students at my university were looking for ways to raise awareness about homelessness in California's Bay Area. Our shared inability to provide the basic necessities of life for every-one who lives in this country was then, and remains today, a significant social and political failure. This is made even worse given the wealth that exists around us in the San Francisco area, a place where so much of the world's wealth is concentrated in the hands of relatively few.

The students had a straightforward goal and their aims were, to put it plainly, good. However, the method they chose to raise awareness wasn't as well-thought through, and their goals and their choice about how to go about raising awareness led to a backlash campaign aimed at getting them to do something (anything) else. For three years, these students would spend a week camping out on campus in tents. Their intention was to visibly live the life of someone experiencing homelessness. To make visible, they thought, what others on campus wished would stay invisible. They did all this, I should say, while still going to their classes, eating in the dining halls, and otherwise still being students when they weren't living in their (expen-sive) tents in a makeshift encampment on one of the campus' well-mani-cured lawns. Some of those involved took a visibly "method" approach to the activity, playacting behaviors they believed really represented what it was like to experience homelessness. These students wanted to experience homelessness and to show others what those experiencing homelessness look like. They wanted to raise awareness and, in doing that, become more aware themselves.

Students don't do this anymore. After several years of criticism from fellow students, faculty, and the community at large, these well-intentioned students began to see why their strategy for raising awareness about home-lessness was misplaced. Many interpreted what the students were doing as a kind of unethical "experience tourism" (of paying money to live like the less fortunate). Others called attention to the fact that these students were rely-ing on stereotypical representations of homelessness when they set up their camps on a university lawn. Still others found it strange that these students did not speak or engage with community activists before deciding on their

DOI: 10.4324/9781003042228-5

choice to create their pseudo-encampment. Is this how members of the community want to be represented? What did they think?

In attempting to raise awareness about homelessness they seemed instead to reinforce harmful stereotypes about the real people they were pretending to be while simultaneously taking attention away from activists doing more practical work on the problem. Thankfully, these students eventually came to realize the error of their ways and shifted their attention toward more productive ways of helping address problems connected to homelessness. I tell this story as a way of getting us to see that many VR and AR developers are engaging in a very similar mistake, although unlike our students, they have yet to realize it.

VR is said to allow for a degree of "homuncular flexibility" (Won, Bailenson, & Lanier 2015). This means that VR seems able to allow for a certain kind of embodied transportation, an illusion of embodiment (the illusion of feeling like a cow or a lobster or another person). Although it's not unique in allowing for this, as the stories above indicate, one of the areas where VR and AR are thought to have significant transformative potential is in the realm of empathy enhancement.[1] Researchers, artists, social activists, and app developers are racing to develop VR and, to a lesser extent, AR simulations that claim to give you the experience of "what it's like" to be any number of things (and I mean this quite literally).

Recall also that VR and AR simulations can create the illusion of "presence" in users. Because of these two features, many researchers, activists, and developers have been trying to harness the power of presence to address social problems, including homelessness. For example, the VR experience *Becoming Homeless* claims to let its users

> spend days in the life of someone who can no longer afford a home. Interact with your environment to attempt to save your home and to protect yourself and your belongings as you walk in another's shoes and face the adversity of living with diminishing resources.
>
> (Ogle, Asher, & Bailenson 2018)

I'll raise questions about whether or not any VR or AR simulations can give you these kinds of experiences (they can't), but the idea that these technologies can function as empathy machines (that they can give you access to someone else's direct experiences) is a tempting one.

To give just one example, in a TED talk given in 2014, Chris Kluwe (former punter for the Minnesota Vikings), argued that AR could change our perception of other people and be used to enhance spectator experience at (American) football games. Kluwe, speaking about how technologies like Google Glass could give football fans a first-person perspective of what it's like to be on the field during a football game, noted that AR technologies can do this "by literally showing someone what it looks like to walk a mile in another person's shoes" (Kluwe 2014). VR and AR advocates are pushing to more widely implement these technologies, not only to enhance entertainment but also to solve pressing real-world problems (homelessness, racism, sexism, ableism, etc.). This is a lot of hope to pin on relatively young technologies.

VR simulations are being created, right now, that promise to show you first-hand what it's like to be a Black man experiencing racist micro and macro aggression. Others say they can help you understand what it's like to become homeless or to experience the symptoms of autism. Still others claim to give you a sense of what it's like to emigrate without documents to the United States from Mexico. More exotic empathy simulations even promise to let you see what it's like to be coral, watching your colony succumb to ocean acidification. The often noble aim of these simulations is to help the user become a better, more empathetic person, to reduce bias, or to better understand the effects of climate change. By providing users an intimate glimpse (from the "inside") of what it's like to experience racism, the developers of these simulations hope to help users change their own thoughts and feelings about race and racism in the future. There's also intriguing early evidence that, at least in the short term, these simulations actually can cause people to change their behavior (Bailenson 2018; Ahn et al. 2016).

This is a chapter on how *not* to use VR and AR technologies. I'll argue that empathy-enhancing simulations like the ones above actually represent a *mis*use of these technologies and that we should not only be seriously skeptical about the power of VR and AR simulation to give us the experience of "what it's like" to be someone (or something) else but also that, in presenting themselves as being able to give us these experiences, that simulations like these are unethical. Empathy-enhancing simulations create several ethical problems for anyone wishing to design or use them. In order to explain why, I'll begin with an extended analysis of these contemporary VR empathy-enhancing simulations.[2]

Once I've shown that such simulations really do promise to give users the experience of what it's like to be something the user is not, I'll show why empathy-enhancing simulations are unethical to design and unethical to use. First I'll argue, harkening back to the arguments in prior chapters, that VR and AR are incapable of giving us the experience of what it's like to be someone (or something else) outside of what I called "base cases" in Chapter 2. Because, by design, all of these empathy-enhancing simulations are guaranteed to *not be* base cases, they are doomed to failure. Second, insofar as they succeed at changing how someone feels, thinks, or acts, then these changes are the result of a distinctly unethical kind of manipulation.

This manipulation really takes several forms, each causing its own unique form of moral harm connected to an ethical subfield known as "nudge" ethics.[3] By presenting users with false experiences, users are manipulated. Additionally, such simulations actually undermine some of the theoretical concepts, like intersectional theories of experience, that they aim to support.

Such simulations will plausibly be judged as unethical from at least four of the most dominant Western moral frameworks in use today. I'll connect these concerns with another problem. Empathy-enhancing simulations face what I'll call the "Intersectional Dilemma." If we are committed to a view about perception that makes it intersectional in nature, a commitment I think is actually quite common, then AR and VR empathy enhancement would violate such a commitment: one or the other must go.

However, while VR and AR technologies are not, and should not be, used as empathy machines they *can* help users become better people in other ways. The problem here isn't with using these technologies for moral enhancement but with *how* we use them to achieve these goals. I'll end the chapter by showing that thinking of VR and AR as sympathy machines can circumvent most of the ethical issues related to empathy-enhancing simulations, and I'll offer practical examples for how we should design and use such sympathy simulations. Here I'll draw from the history of anti-bias technologies to show that VR and AR are arguably the best anti-bias technologies we have available so long as we design those simulations to generate sympathy. There's a great store of currently untapped potential in VR and AR as anti-bias, pro-social tools.

VR and AR as Empathy Machines

In Chapter 2 I talked about the limitations of the imagination in the context of the Thought Experiment Paradigm. There I noted that VR and AR technologies hold a lot of promise as tools for helping us engage in at least some kinds of "in-their-shoes" empathizing. In-their-shoes empathy, in this context, means that these technologies can, in well-designed simulations, give users a sense of how they might act under certain circumstances.[4] In Chapter 4 we looked at some of the *moral* risks that come with making and using well-designed simulations like these. Given these features of VR and AR, it might not be all that surprising that scholars, activists, and app designers have designed simulations that only promise to get us to experience "in-their-shoes" empathizing in ways that, I'll argue, stretch these technologies beyond what they're capable of.

Take, for example, a simulation called *1,000 Cut Journey*. The creators of this simulation promise that, in using it, "the viewer *becomes* Michael Sterling, a Black man, encountering racism as a young child, adolescent, and young adult" (Cogburn et al. 2018). Courtney Cogburn, chief creator of *1,000 Cut Journey* and Associate Professor of Social Work at Columbia University, has claimed that, in using her simulation,

> it seems that white people who go through the experience are moved. Tears are not uncommon. They seem to connect more deeply with something they know and understand conceptually but have never felt in quite the way that happens in the VR experience.
>
> (Grimsley-Vaz 2018)

In order to make sense of Cogburn's claims, we need to better understand the underlying assumptions that ground them. One assumption here is that VR technologies in general, and simulations like *1,000 Cut Journey* in particular, are giving their users a unique perspective that they didn't have before. The simulation is presumed to give people direct, first-person, access to the experience of "what it's like" to be someone they aren't. Cogburn's White subjects, by her own account, might have understood racism "conceptually" but did not understand it *personally* until they had gone through the simulation. The simulation, in other words, now showed them what it was like to experience anti-Black racism. This set of assumptions is common. Developers of empathy-enhancing VR simulations all seem to accept something along these lines, and it's worth highlighting it now and keeping it in mind.

I Am a Man is a VR simulation with similar goals. Created by Derek Ham, professor of graphic design at North Carolina State University, the simulation makes similar promises to its users about the kind of empathic access it can give them to the inner life of a simulated Black man. Ham claims that "[t]he VR experience allows one to literally walk in the shoes of people who fought for freedom and equality during the civil rights era" (Ham 2018). The same assumptions that ground Cogburn's simulation also ground Ham's simulation.

Another example (because it matters that these assumptions are widespread and worthy of critique): Director Alejandro González Iñárritu created a travelling VR museum exhibition called *Carne y Arena* ("Flesh and Sand"). In this simulation, which makes impressive use of room-scale VR technologies, users go through the experience of migrating to the United States from its southern border. They take on the role of an undocumented immigrant travelling in a group with others as they attempt to evade immigration officials and other threats. Iñárritu has said that his

> intention was to experiment with VR technology to explore the human condition in an attempt to break the dictatorship of the frame, within which things are just observed, and claim the space to allow the visitor to go through a direct experience walking in the immigrants' feet, under their skin, and into their hearts.
>
> (Iñárritu 2017)

VR developers are currently racing to create ever more of these empathy-enhancing simulations. Iñárritu too appears to think that his simulation can give users special access to the experiences of other people.

Simulations aren't limited to giving people access to the experiences of people from different racial categories but also of those who are differently abled. A company called Viscira, for example, created a VR simulation designed to give people insight into what it's like to have schizophrenia (Couch 2016). Other simulations exist designed to give users the experience of being pregnant, wheelchair-bound, and so on. The creators of *Autismity: The Autism Simulator* say that its users will "experience with full intensity how hypersensitivity, sensory overload, and synesthesia feel. The ability to experience sounds, lights, patterns, and disturbing elements, as autistic individuals do, shapes the revealing insight into their daily experiences" (Autismity 2016).

Researchers investigating the limits of homuncular flexibility have even devised experiments that, they assume, can help users better empathize with non-human animals like cows. Sun Joo Ahn and colleagues (2016), for example, created a simulation that aimed to provide users the experience of what it's like to be a cow being raised for slaughter. During the study, "participants got down on their hands and knees, and embodied a virtual shorthorn cow in a pasture where they saw their cow avatar directly facing them as if looking into a mirror" (404).[5] Ahn et al. are, like the designers of other VR empathy simulations, operating under the belief that

> [b]ecause IVEs [immersive virtual environments] allow individuals to put themselves inside the virtual body of an animal, they would *directly feel* the threats it is up against and feel *connected* to its plight. For instance, *sharing the experience* of body transfer of oneself to the cow's virtual body would clearly help people *understand how a cow would feel* being raised for its meat.
>
> (Ahn et al. 2016)

The language here is important. Ahn et al. seem to work under the assumption that their simulation can, literally, "help people understand how a cow would feel" and that this new understanding will shape their future choices about what to eat. We saw, with Kluwe's case, that this assumption is true for designers of both VR and AR empathy-enhancing simulations. In most cases, the intentions of the creators of simulations like these, as we saw with *1,000 Cut Journey*, are noble: to make users more aware of animal welfare or to help them become less racist, xenophobic, or ableist. They do this via the supposed ability of VR and AR to let us see, directly, what it's like to be a different sort of person and to experience their hardships.

What makes these simulations "empathy-enhancing"? The argument that grounds these simulations goes something like this:

1 One of the major sources of racism (or sexism, ableism, xenophobia, etc.) is a lack of empathy
2 Virtual reality is an empathy machine
3 Because of this, some VR simulations can give users empathic access to the experiences of others
4 Users of these simulations seem to genuinely change after experiencing them
5 If VR is an empathy machine, then using VR can be an important way to remove racism (or sexism, ableism, xenophobia, etc.)

It's hard to make sense of VR simulations like *I Am a Man* or *1,000 Cut Journey* or *Carne y Arena* without believing that their designers subscribe to most of these claims. The key assumption, in my view, is the third one. If VR, or AR, really can give us access to the experiences of others (what we called in Chapter 2 "simulation" empathy) then the rest of the premises follow logically. But VR and AR can't allow us to engage in simulation empathy except in what we called base cases. The assumption that experiences can be transferred via VR and AR simulation falls prey to a problem.

The problem I'm posing is what some philosophers of technology have called the problem of semantic variance. Philosopher Trevor Pinch, following philosophers like Evan Selinger and Kyle Whyte (Pinch 2010; Selinger & Whyte 2010), define the problem of semantic variance this way:

> Semantic variance simply expresses the point known to all interpretive sociologists that meaning depends upon context. The problem of semantic variance ... is not just a philosopher's way of posing a counter example. It goes to the heart of the issue of how people routinely interact with technology and points to the wider social and cultural framework within which technology is embedded. No theory ... will be satisfactory without taking on board these sorts of factors.
>
> (Pinch 2010)

For empathy-enhancing VR nudgers, the problem of semantic variance relates to whether or not VR and AR experiences are able to translate the semantic elements of experience from one person to another. In other words, can they make good on their promise to give their users the experience of "what it's like" to be someone (or in the case of Ahn et al.'s slaughterhouse simulation, of *something*) else. Empathy-enhancing simulations only work if the experiences they give their users have semantically invariant content (content that doesn't change from user to user). I'll argue in the following sections that this is almost certainly a false assumption and that because of this, empathy-enhancing simulations can't deliver on their promises. VR and AR simulations can't give us access to the inner lives of other people in non-base cases for the same reasons we discussed in Chapter 2 when we discussed the problems embedded within the Thought Experiment Paradigm.

Base cases, recall, were cases where the differences between people were irrelevant to the outcome. If two people are playing chess (to win) and player A is one move away from a checkmate then, in that limited case, we can all empathize with A in both "in-their-shoes" and "simulation" senses of empathy. Assuming that we know the basic rules of chess and understand the current position of the pieces on the board, we should all be able to see what the next move is going to be. However, none of the simulations we're discussing now are base cases. They depend, by design, on users being different from the people whose lives are being simulated in order to have their effects. As such, we run into a variation of the problems we raised in Chapter 2 about the different roles that doxastic (conscious) and subdoxastic (non-conscious) elements play in determining our experience.

Philosopher Thomas Nagel (1974) produced a now famous argument with a very similar conclusion. In his classic "What is it like to be a bat?" Nagel essentially argued that it would be impossible for us to imagine what bat experiences are like because bat experiences are shaped by bat bodies (by the distinctive way that bats are embodied) and that this is different enough from human embodiment to mean that the task of imagining what it's like to be a bat, for a human, is a hopeless one:

> It will not help to try to imagine that one has webbing on one's arms, which enables one to fly around at dusk and dawn catching insects in one's mouth; that one has very poor vision, and perceives the surrounding world by a system of reflected high-frequency sound signals; and that one spends the day hanging upside down by one's feet in an attic. In so far as I can imagine this (which is not very far), it tells me only what it would be like for me to behave as a bat behaves. But that is not the question. I want to know what it is like for a bat to be a bat. Yet if I try to imagine this, I am restricted to the resources of my own mind, and those resources are inadequate to the task. I cannot perform it either by imagining additions to my present experience, or by imagining segments gradually subtracted from it, or by imagining some combination of additions, subtractions, and modification. To the extent that I could look and behave like a wasp or a bat without changing my fundamental structure, my experiences would not be anything like the experiences of those animals.
>
> (Nagel 1974)

I quoted philosopher Peter Goldie in Chapter 2 arguing for a similar point:

> A cannot, as part of a consciously willed project, keep Bs characterization in the non-conscious background in her imaginative exercise of wondering what B will decide to do in a certain situation. A will be obliged, in trying to shift to Bs perspective, to treat Bs characterization through the theoretical or empirical stance, as one typically does when considering the role of character in explaining or predicting other peoples' decisions, actions, and so on ... this produces a fundamentally distorted model of Bs thinking.
>
> (Goldie 2011)

Nagel and Goldie are in agreement here, I think, about the impossibility of experiencing what it's like to be a bat. However, we might imagine that developers of VR and AR empathy-enhancing simulations might have a response to this kind of argument.

VR and AR simulations, they might argue, have an advantage over mere imagination in that they offload the imaginative task onto virtual environments and so can escape the problems laid out by Nagel and Goldie. Could Ahn and colleagues be right about the power of VR to let us have direct access to cow experiences? Empathy-enhancing VR simulations assume that what it means *to be* a cow, or Michael Sterling, or experience autism, can be given to users by holding a (360-degree) virtual camera to the events in their lives. That such a point of view will capture and transfer his experiences to the viewer as these individuals experienced them. Further, they assume that it's morally (and, as we'll see, metaphysically) unproblematic to change user behavior on the basis of these experiences. I think both assumptions are false. To see why we need to revisit the issue of experience, first explored in the second chapter, and say more about what it means to have an intersectional theory of experience. If experience is intersectional in any way, then VR and AR empathy-enhancing simulations are bound to fail due to what I've called the Intersectional Dilemma.

Furthermore, if such simulations are bound to fail but are nonetheless used to alter user behavior (even for noble causes), then they are almost certainly unethical to use. It will be, I'll argue, unethical for people to develop or use them.

Intersectionality as a Problem for Empathy-enhancing Simulations

The problem of semantic variance suggests that in order for empathy-enhancing VR simulations to successfully transfer the experiences of one person to another, the vehicle for that transfer (the VR technologies themselves) must not alter the content of the experience.

Our experiences are almost always *about* something (I'm looking at a cute puppy, I'm smelling the pizza roasting, I'm feeling my leg slowly get numb, etc.). In Chapter 2, we spoke about how the content of an experience is partially composed of both conscious and non-conscious elements. The *aboutness* of experience (i.e., what it is that our experiences are directed at) is usually conscious (we consciously think or see specific things) but these can be influenced by non-conscious elements.

When critiquing the Thought Experiment Paradigm, for example, I said that some of the elements that impact our conscious reactions to thought experiments (like the trolley or ticking-time-bomb cases) are influenced by non-conscious elements. For example, the anxiety or stress most of us would feel by being under time-pressure during a moral dilemma can influence how we would experience the actual event (how we would process the sights, sounds, smells, and feels of the moment, what sorts of features in the environment itself would get most of our attention, and so on). Near the end of that chapter, I argued that VR and AR might actually make it possible (though maybe unethical) to engage in specific kinds of "in-their-shoes" perspective-taking.

However, empathy-enhancing VR and AR simulations claim to be able to not only transport *you* to new worlds to experience discrimination, homelessness, and so on, they claim to give you the experiences of other people who are very different from you in at least some way. I provided direct accounts from the developers themselves, to show that they promise to let you experience what it is like for someone else to have these experiences, from their point of view. In this section, I consider one major problem with this kind of empathy-enhancing simulation: it goes against an important (but of course not the only) tenet in many models of experience: that the content of our experience (what it is that our experiences are about) is intersectional.[6] In the section above, I mentioned Thomas Nagels' arguments about the limits of the imagination to simulate bat experiences. Although he wouldn't use this term, Nagel's argument hinges on what I'll be calling a theory of experience grounded on a commitment to structural intersectionality.

The term intersectionality itself was first coined by Kimberlé Crenshaw (1989; 1991) though the concept can be traced back to early work by Black feminist activists (Combahee 2015). In the intervening decades, this term has entered popular American and global culture, and captured the imagination. It has, unsurprisingly, taken on many different meanings. Because of this, it's important for us to clarify how I'll be using the term intersectionality and the various senses of the term. Crenshaw herself set out to try and describe how different aspects of our identities (i.e., race, class, sex, gender, orientation, nationality, geography, etc.) can affect important elements of our lives in complex ways that can be hard to predict in advance.

In these early works, Crenshaw distinguished between three different ways that she thought aspects of a person's identity could intersect and referred to these different forms of intersectionality as structural, political, or representational. Each has given rise to its own form of intersectional analysis or theory. Although the form of intersectional theory I'll be focusing on in my critique of empathy-enhancing VR and AR simulations comes closest to what Crenshaw called "structural" intersectionality, it's worthwhile to say what these distinctions amount to and how they're connected to the problem of semantic variance and Nagel's earlier point about bats.

Intersectionality is probably best known in connection to theories of oppression and power. Crenshaw referred to this kind of intersectionality as "political" intersectionality. Political intersectionality is meant to help us explain how gender and race, for example, can combine (can "intersect") to create forms of oppression that are not reducible to gender discrimination or racial discrimination:

> [i]t is not just that [Black women] have quantitatively greater hurdles to overcome than white women or black men, but that the nature of their oppression reflects a distinctive, complex, and perhaps irreducible

combination of sexism, racism, and other structures of oppression, such as classism and heterosexism.

(Gasdaglis & Madva 2019)

It's reductive, on this view of intersectionality, to talk simply about gender discrimination without saying more about the full context in which that discrimination takes place. Through the lens of political intersectionality, the way that oppression works against women of different races or ethnicities isn't straightforwardly comparable (they can't be summed up against one another easily). A wealthy native-born White woman will have misogynist structures operate against her in a different way than a wealthy native-born Black woman. Race, gender, class, and the like "intersect" in ways that change how oppression can impact individuals (even if the oppression appears to be purely about gender). Similarly the forms of oppression operating on Black citizens in the US can't be neatly compared against other forms of oppression without taking into account how intersecting axes of identity shape that oppression (how wealth, gender, and so on affect how that oppression looks and feels).

Like political theories of intersectionality, representational theories of intersectionality look at the way that cultures represent identity. "Representational intersectionality concerns the production of images of women of color drawing on sexist and racist narrative tropes, as well as the ways that critiques of these representations marginalize or reproduce the objectification of women of color" (Carastathis 2014, 307). It's important that we appreciate both the similarities and differences between these types of intersectional theories. Although they differ in terms of their focus (oppression versus forms of representation) they share a commitment to viewing elements of identity as interacting with one another in complex ways to produce a more subtle, unique, understanding of an individual. The same is true of "structural" theories of intersectionality. A structural intersectional analysis focuses directly on explaining the content of a person's experience. In other words, structural intersectional theories look at how identity can impact the basic nature of individual experience and how that might vary from person to person. Philosopher Sara Bernstein puts the point this way:

[Structural] intersectionality … refers to a type or token of experience faced by members of [intersecting identity] categories, as in experiences had by black women that are not entirely explicable by appeal to being black or to being a woman.

(Bernstein 2020)

In one of her early articles on the subject, Crenshaw (1991) discussed the concept of "structural intersectionality" in terms of experience as well. According to Crenshaw, structural intersectionality, as a framework or theory, can help us understand how the location of women of color at the intersection of race and gender makes [their] actual *experience* of domestic violence, rape, and remedial reform qualitatively different than that of white women.

(Crenshaw 1991)[7]

As yet further evidence of this way of thinking about structural intersectionality and its relationship to experience, philosopher Elena Ruíz (2017) has also discussed intersectionality largely as a theory about the content of a person's experience.

As a descriptive term, [intersectionality] refers to the ways human identity is shaped by multiple social vectors and overlapping identity categories (such as sex, race, class) that may not be readily visible in single-axis formulations of identity, but which are taken to be integral to robustly capture the multifaceted nature of human experience.

(335)

Thus, while intersectionality is itself a complex concept with its own multifaceted meanings, I've tried to suggest here that intersectional theories of experience (i.e., "structural intersectionality") have many adherents.

An intersectional theory of experience, then, is one in which the content of an individual person's experience, at any given time, is going to be shaped not only by where they happen to be looking, but also by the effects that their internalized concepts of race, gender, class, nationality, etc., have on those experiences.[8] This will be true at both the doxastic and subdoxastic levels as our internalized concepts of race, gender, class, ability, and so on exercise top-down effects on how we make sense of the world and symbols around us. With this terminology in mind, we can get back to Nagel.

Nagel (1974) argued that some people might believe that it really is possible to understand what it's like to be a bat if they tried really hard to imagine it. To do so, we would have to try and escape the subjectivity of the bat's perspective and try to describe bat minds, assuming they exist, in more objective terms. Nagel argued that

[i]t is difficult to understand what could be meant by the objective character of an experience, apart from the particular point of view from which its subject apprehends it. After all, what would be left of what it was like to be a bat if one removed the viewpoint of the bat?

(443)

Nagel expresses, throughout his article, a commitment to experiences having a certain kind of subjective character. This subjectivity is, in his view, related intimately to the way that an animal evolved and, beyond that, with the specific capacities it has.[9] It's in this sense that Nagel's theory of subjective experience is committed to a form of structural intersectionality. I'm spending some time on this point to clarify that a commitment to structural intersectionality may, but doesn't require, a commitment to other forms of intersectionality and also that structural intersectionality is a common, though of course not universal, commitment for scholars of mind and perception. It will take a good deal of evidence, in other words, to show that our experiences are transferable via VR and AR if experiences can be even partially understood through the lens of structural intersectionality.

The problem of semantic variance now really threatens the viability of empathy-enhancing VR and AR simulations by posing an Intersectional Dilemma. If the semantic content of our experiences is even partially determined by something like structural intersectionality (that is, by the top-down effects of internalized concepts of identity) then it's hard to see how any VR simulation can give us the sort of direct access to the experiences of other people no matter how well designed. Either structural intersectionality is false (a commitment likely to go against the beliefs of many developers of VR and AR empathy-enhancing simulations) or empathy-enhancing simulations of the kind I've been discussing are impossible.

Goldie's (2011) argument about empathy failures when dealing with non-base cases is poised to strike again here. None of the empathy-enhancing simulations I've discussed in this chapter aim to recreate base cases (base cases would be beside the point of the simulations as they would no longer be enhancing user empathy but only mirroring user perspectives). Simulations like *Becoming Homeless* are meant to be experienced by users who haven't personally been affected by homelessness. *Autismity* is intended to be used by people *not* on the spectrum. *Carne y Arena* targets users who are not themselves undocumented. Because of this, they're bound to fail at transferring experiences.

Much as we said in Chapter 2, empathy-enhancing simulations might be able to give us some sense of what it might be like *for us* to undergo similar sorts of experiences, but *my* experience of simulated pregnancy isn't likely to mirror the experiences of an actually pregnant person. My experience of walking on all fours and being poked in the sides, as I might be in Ahn et al.'s slaughterhouse simulation, is unlikely to mirror a shorthorn cow's experiences of the same events – indeed how could it? Though as of this writing a VR or AR bat simulator doesn't exist, it seems likely that Nagel would agree that such a simulation couldn't possibly give us the experience of what it's like to be a bat but only the experience of a person trying to be a bat.

Equally problematic is that, in attempting to recreate the experiences of people, these simulations go against the widely shared commitment that experience is affected by structural intersectionality. Michael Sterling's experiences aren't determined by whatever a neutral camera records. Sterling's experiences are given meaning by the intersecting effects of his own self concepts (as a Black man living where and when he lives). If I, a male Nicaraguan immigrant to the US who grew up in poverty largely in Southern

California in the late 20th century, were to put myself through *Preterna* (Thapan 2016), a VR pregnancy empathy simulation, I might be able to learn something about what it would mean *for me* to suddenly have a pregnancy belly (though even here, I wouldn't learn anything about what this would *feel* like – the haptics aren't there yet) but I wouldn't be any closer to learning what it's like for someone to go through an actual pregnancy.

In part this would be because what it would be like for someone to become pregnant drastically depends, unsurprisingly, on *who* they are, whether the pregnancy was planned, their age, whether the pregnancy is still desired, and so on. Any real-life situation involving someone becoming pregnant would be affected by these features and many of these effects would be non-conscious (subdoxastic). For all the same reasons Goldie (2011) argued that we can't consciously simulate these non-conscious features using our imaginations, we can't consciously fill these intersectional elements in with a VR or AR simulation. These technologies can't transfer experiences of this kind without robbing them of their essential natures.[10] The Intersectional Dilemma looms too large.

We have some evidence for thinking that our internalized concepts can play a role in how we experience or perceive the world and thus some evidence to support structural theories of intersectionality. For example, Alessio Avenanti and his colleagues investigated whether internalized in-group and out-group biases affected the way that people processed pain (Avenanti, Sirigu, & Aglioti 2010). Their group showed subjects images of hands with different racial markers. Hands were selected that represented either White or Black subjects. Researchers also included images of a non-racialized violet hand. In all cases, each hand was subjected to a painful stimulus (a needle prick).[11] Avenanti et al. were interested in seeing whether the subject's race affected their empathic suffering when viewing another person feeling pain. Their results have interesting implications for VR and AR empathy simulations. According to the researchers, their

> results support the notion that perceiving bodily stimulations on ingroup members leads to an immediate resonance with affective and sensorimotor components of the observed feelings. In contrast, responses to outgroup members' somatic stimulations are less embodied and automatic and likely rely more on slower controlled processing.
>
> (Avenanti, Sirigu, & Aglioti 2010)

Empathic contagion appeared, in this study, to be affected by unconscious judgments about whether the hand in question was seen as being in-group or out-group, and these concepts also appeared affected by internalized ideas about race in connection with in/out-group membership:

> a clear sensorimotor contagion was found not only in response to the pain of stranger individuals belonging to the same racial group but also

in response to the pain of stranger, very unfamiliar but not culturally grouped, individuals (violet models). By contrast, *no* sensorimotor contagion was found in response to the pain of individuals culturally marked as outgroup on the basis of the color of the skin of a nonfacial body part that did not express any specific emotion.[12]

(Avenanti, Sirigu, & Aglioti 2010)

Another, now famous, series of studies investigated whether internalized racial concepts affected how people disambiguate ambiguous objects in a "shoot/don't shoot" police simulation (Correll et al. 2002). Subjects were shown photos of men whose racial characteristics were varied so as to be read as being White or African American. Each man in the photo was holding something in his hands which, when viewed quickly, was intentionally designed to be somewhat ambiguous. The key question for Joshua Correll and his colleagues was whether racial characteristics would impact how subjects read the ambiguous objects in the men's hands.

The results of these studies consistently support the hypothesized effect of ethnicity on shoot/don't shoot decisions. Both in speed and accuracy, the decision to fire on an armed target was facilitated when that target was African American, whereas the decision not to shoot an unarmed target was facilitated when that target was White.

(Correll et al. 2002)

As we'll see further below, these findings are supported by research on bias reduction and pain perception across cultures (Sheng et al. 2014; Xiao et al. 2016). Even seemingly unconscious experiences, like mirroring empathy (i.e., empathic distress), appear to be affected by the internalized racial concepts that we bring to an experience. If something as simple as the activity of mirror neurons in the sensorimotor cortex can be influenced by internalized racial concepts, then we have some additional empirical support to think that structural intersectionality is true for at least some forms of experience, and this spells trouble for empathy-enhancing simulations.

One recent meta-analysis of studies examining VR empathy-enhancement simulations (of all sorts) conducted by Alison Martingano, Fernanda Herrera, and Sara Konrath (2021) is consistent with the arguments I'm making here. Martingano and her colleagues conclude that

VR improved emotional empathy, but not cognitive empathy. In other words, VR can arouse compassionate feelings but does not appear to encourage users to imagine other peoples' perspectives. Further subgroup analyses revealed that VR was no more effective at increasing empathy than less technologically advanced empathy interventions such as reading about others and imagining their experiences.

(Martingano, Herrera, & Konrath 2021)

In the next section, I lay out some reasons to think that developing and using empathy-enhancing simulations like these is unethical. My arguments will depend in part on the problem of semantic variance being insurmountable for non base-cases. At the end of the chapter, I'll consider the possibility that sympathy-enhancing simulations could sidestep most of the problems that I raise for empathy-enhancing simulations. The emotional empathy identified by Martingano and her colleagues will play a role in the development of these sympathy simulations. It is sympathy, not empathy, that is the true moral domain for VR and AR bias reduction.

Empathy-nudging: Some Problems

Nudges are manipulations in an environment, real or virtual, whose intention is to influence (but not determine) the behavior of people in that environment. As philosopher Cass Sunstein puts it: "[n]udges are interventions that steer people in a particular direction while still allowing them to go their own way" (Sunstein 2015). Because of the potential for nudges to have large-scale social effects, the ethics of nudging has received a lot of attention from philosophers and other researchers (Bovens 2009; Sunstein 2015; Hansen & Jespersen, 2013).

In this section, I'll briefly summarize the nature and ethics of nudging to show that VR empathy-enhancing simulations are unethical because they violate the restrictions on morally acceptable nudges. It's clear, from the preceding sections, that the designers of empathy-enhancing VR simulations *intend* for them to function as nudges rather than as entertainment so it makes good sense to see if the literature on the ethics of nudges can help us make sense of why it might be wrong to design and use these simulations.

The ethics of nudging is normally understood to revolve around three ethical concepts: transparency, autonomy, and welfare (Bovens 2009). These three concepts are interrelated, as we'll see, but it's useful to look at them separately. Transparency, in the case of nudges, means that the nudge itself ought to be visible to a careful observer. Calorie information on a menu, for example, is a transparent nudge. A person ordering food from a menu with calorie information printed on it can reasonably infer that that information is there to inform (and influence) their choices about what to order. Pictures of diseased organs sometimes placed onto packs of cigarettes are also transparently placed so as to make the nudge obvious to users. They are there to influence, but not determine, a smoker's choice to smoke. Note also that all nudges are manipulative and that this isn't always ethically problematic. Manipulation, in this case, refers to anything that aims to non-rationally change a user's thoughts, behaviors, or feelings. Some ways of being manipulated are benign (e.g., a speed bump manipulates most drivers into slowing down) and some ways of being manipulative are not so benign (Noggle 2018). The more transparent, truthful, and aligned with a person's interests a manipulation is, the less likely we are to think it ethically objectionable.[13]

Importantly, transparent nudges like the two mentioned above do not lose their power as nudges just because they're transparent. You know exactly why that calorie information is there and what you're meant to do with it. Those photos on the cigarette packs are not there to make you feel good about your decision to smoke. In both cases, the user can choose to disregard the nudge but will be hard pressed to ignore it altogether. Transparency is important in an ethical analysis of a nudge because more transparent nudges are less likely to be unethically manipulative or coercive (coercive nudges are not nudges at all). The more opaque the nudge, the greater the ethical justification required for it.

Transparency is also important because transparent nudges are less likely to subvert a person's autonomy. We care about our ability to plan out our lives, to act for reasons, and to choose what we value and how to best go about valuing it. Because nudges are meant to influence this process, it's important to think about whether or not nudges are intended to *help us* achieve our goals or to influence us to do (or care) about things that we wouldn't otherwise care about. We may, for a lot of reasons, care about calories and so might be happy to have calorie information on our menus. On the other hand, perhaps we don't care too much about the increased health risks associated with smoking and find photos of diseased organs distasteful. Because these nudges are transparent, they aren't unethical to use even in cases where the user chooses to neglect the nudge. A speed bump is a nudge to slow down while driving (and a pretty transparent one at that!) but here too users can choose to ignore the nudge and speed through the bump. Transparent nudges face fewer ethical hurdles in part because they make it harder for us to be unethically manipulated via deception. A transparent nudge thus may remain ethical even if it's nudging us toward things we don't want or even toward things we find immoral, within limits.

Nudges are easiest to justify when they're transparent, don't subvert our autonomy, and improve welfare. Welfare, in these cases, might refer both to the welfare of the individual nudgee or to social welfare more generally.[14] Think back to the images of diseased organs. Whether or not that nudge advances welfare will depend on several factors, some of which might be contradictory. For example, does the smoker care about their health? How much do they value the pleasure of smoking? Additionally, social welfare questions are relevant too: is society better off, all things considered, with fewer people smoking? Transparent nudges that respect individual autonomy and that are aimed at advancing user or social welfare (ideally both!) are thus the least problematic kinds of nudges, morally speaking. Nudges that are less transparent (or that rely on being opaque to work), coercive, or which harm individuals or social welfare thus require ever increasing justification to use.

Before turning back to thinking of empathy-enhancing VR and AR nudges, it's important to look at one more complication that comes up in the context of the ethics of nudging. While all nudges aim to influence users, some nudges aim to influence users specifically by educating them. These "educative" nudges "attempt to inform people, so that they can make better choices for themselves" (Sunstein 2015, 427). Because empathy-enhancing nudgers aim to give users information that's meant to help them make better choices, they deserve extra ethical scrutiny about the information they're conveying.

Calorie information on a menu is not only a nudge, it's an educative nudge. In order for nudges like these to be ethical, they must give people accurate information even if doing that makes them less effective as nudges.[15] Not all nudges are educative. Speed bumps, the layout of a grocery store, high energy music in a gym, and so on are all there to influence users toward making specific choices but they don't do this by giving their users extra information. They're all nudges, but not educative ones. Educative nudges, because they rely on the force of the information they give to us, must be held to an especially strict standard if they're to be ethical to use.

Imagine if, for example, a written warning was printed alongside images of diseased organs on packs of cigarettes and that the warning vastly misrepresented the risks associated with smoking. While such a nudge would, in one sense, be transparent (it's clear what it's doing printed on the pack) and while it may (arguably) be in our interest to be nudged into non-smoking, such a nudge may yet be unethical on the basis of providing intentionally misleading information to users and thus, as an educative nudge, violate the autonomy of those nudged by it and thus be unethical to design or use.[16]

It seems clear that VR and AR empathy-enhancing simulations are intended to function as educative nudges. For such simulations to work, they must be able to accurately convey information ("what it's like" to be the kind of person or thing the simulation is about) to users. It's also clear that they intentionally represent themselves as giving users this kind of information. Users are then meant to take the information given to them by the simulation ("Oh! That's what it's like!") and to use it when they make future decisions.

When we assess the ethics of nudges, it's also important that we stay focused on the actual mechanics of the nudge itself and not what the nudge is getting us *to do*. It's entirely possible, as I think is the case with many empathy-enhancing VR and AR nudges, that an unethical nudge may have morally good intentions and goals. Research suggests that all of us tend to forget this distinction and that we approve of nudges whose goals we agree with and disapprove of nudges whose goals we disagree with:

> both laypeople and practising policymakers evaluate policy nudges in ways that are coloured by their political preferences. People tend to view nudges as more unethical, coercive and manipulative when illustrated by policy objectives they oppose compared with objectives they support (study 1), or when told that such behavioural interventions have been enforced by a policymaker they oppose compared with one they support.
>
> (Tannenbaum, Fox, & Rogers 2017)

This is important, on the one hand, because we want to focus on the ethics of the nudge itself and not necessarily what the nudge is trying *to do* to us. Aldous Huxley's *Brave New World* (1932/2010) may show us a picture of a world where most people are content with their lives, a laudable moral goal, but at the same time it shows us that the process required to achieve these ends was deceptive, lacked transparency, subverted autonomy, and thus was unethically manipulative. Most scholars who assess the ethics of nudges choose, for these reasons, to focus their assessments on a nudge's structure (whether it is transparent, respective of autonomy, and whether it promotes individual or social welfare) instead of its goals (Sunstein 2015; Blumenthal-Barby 2013; Hansen & Jespersen 2013).

When discussing the welfare component of nudges, I mentioned that both individual *and* social welfare should factor into that assessment. The social goals of empathy-enhancing nudges aren't irrelevant to our moral calculus (good/bad? permissible/impermissible?). On balance, I'll argue that these nudges are unethical for anyone to use or develop and that they're unethical for two important reasons. First, their force as nudges depend almost entirely on their lack of transparency and inaccuracy (as educative nudges this matters a lot). Second, and as a result, they're deceptively manipulative and thus subvert the autonomy of anyone who uses them.

Still, it might be argued that the goals of these nudges (creating a less racist, less ableist, less sexist, more empathetic society) are so good that the ends justify the means. I appreciate the force of this objection. In what follows, I'll suggest that these nudges are not only likely *not* to actually succeed at their goals but that other, more ethically permissible, kinds of simulations stand a better chance of doing all of these things. If I'm right, then many empathy-enhancing VR nudges are both fruitless and unethical. Let's look at these arguments.

Why think that empathy VR nudges lack transparency? After all, one might argue, they are almost always very clear about their intentions. In some cases, as with *Becoming Homeless: A Human Experience* (Ogle, Asher & Bailenson, 2018) the title itself seems to transparently convey the purpose of the simulation and its intentions as a nudge. One reason to think that these simulations are nonetheless opaque is connected to what I've been arguing throughout this chapter: that they aren't able to deliver on the promise to transfer experience.

When Derek Ham, creator of *I Am a Man*, says that "[t]he VR experience allows one to literally walk in the shoes of people who fought for freedom and equality during the civil rights era" (Ham 2018), he misrepresents what the simulation is capable of doing. Users are not literally able to walk in the shoes of people who fought for freedom and even less literally are they able to walk in the shoes of the Black male subject whom they embody in that simulation. The actual force of empathy-enhancing VR nudges is thus dependent largely on having users believe that they're having a certain kind of experience that they're *not really having*. We should worry about the brittleness of nudges like these *even if* they turned out to change behavior in the short term. Much like the cigarette nudge that radically overstates the risk of smoking, once people find out about the nudge's lie not only is the nudge itself likely to lose its effectiveness but it's also likely to undermine people's confidence more generally about medical warnings.

While it's partially an empirical question whether people would still be nudged by these simulations if designers were honest about the limitations of their simulations, it seems likely, at least to me, that these simulations would lose much (maybe most) of their nudging power by being transparent in this way ("this simulation won't show you what's it like to experience racism, but it will show you what it feels like to have people treat you as if you were someone you're not" has a lot less impact). Similarly, a simulation like *Becoming Homeless*, if made transparent, would appear almost self-defeating ("this isn't actually a simulation about becoming homeless but maybe you'll learn something"). So we have some reason to think that these seemingly transparent nudges are not transparent at all.

As educative nudges, VR and AR empathy simulations are also problematic. Because they cannot, except in the rare instance of the base case, actually give users the information they claim to give, they depend on delivering false information for their nudge power. While a menu with highly inflated calorie information might be a more effective nudge than a menu with accurate calorie counts (if the goal is reducing total calories consumed), such a nudge would not only cease to be educative (because users are not actually being educated) but would also be unethically manipulative. Empathy-enhancing nudges run into the same problem. In deceiving users about the nature of their experiences, they subvert user autonomy. If the ethics of nudging is dependent on a combination of transparency, respecting autonomy, and enhancing individual or social welfare, empathy-enhancing simulations run into serious ethical problems.

Before moving on, I would like to add two more problems for VR and AR empathy-enhancing simulations. Imagine a "best case" scenario for an empathy-enhancing VR nudge. Suppose that I were to enter a simulation that was a base case for me (it simulates some of my own experiences growing up as a refugee and migrant in Los Angeles county in the 1980s and 1990s), and suppose further that I was successfully nudged by this simulation into changing some of my beliefs. Some philosophers have argued that even in cases like this, ethical issues remain. Philosopher John Harris, for example, has argued that

> [i]t seems to me that moral enhancement, properly so called, must not only make the doing of good or right actions more probable and the doing of bad ones less likely, but must also include the understanding of what constitutes right and wrong action. As I have suggested elsewhere, there is no virtue in doing what you must.
>
> (Harris 2013)

Harris' concern here might be that most nudges, even when they're successful, may "merely" nudge users into compliance and that this isn't enough for someone to actually have changed for the better. For philosophers like Harris, doing the right thing requires that someone knows what the right thing to do is (that is, they understand morality well enough to get at the moral right answers consciously) and then also have the strength of will and conviction to actually go on to do it. Even educative nudges might not be morally praiseworthy for someone like Harris.[17]

As I'm writing this, citizens in the USA are nearing the end of another bitterly contested election season, perhaps the most bitterly contested in our lifetimes. Imagine that someone were to get a pamphlet that tells them how their political party recommends they vote. These pamphlets are clearly educative nudgers (they work because they're giving users information meant to affect their decisions without being coercive). Imagine further that our hypothetical voter then simply votes exactly as the pamphlet tells her to. For someone like Harris, this is not yet enough to have made the voter's choices morally good because the voter herself had a relatively small role to play in making her decisions. She isn't taking full ownership over her choices and it's not clear, to philosophers like Harris, that her actions deserve any moral praise.[18]

Concerns like these might affect some of the educative nudges I've been discussing in this chapter. If empathy-enhancing nudges deliver false information then they're unethical for being deceptively manipulative and opaque. On the other hand, even if I'm wrong about the possibility of transferring first-personal experiences via VR or AR, it's not clear even in these cases that empathy nudges would be morally good, or at least that they wouldn't be morally preferable to having someone reach these conclusions on their own without the manipulation caused by being nudged.

But we have yet another set of concerns that threaten VR empathy nudgers like this. Even if it turns out to be possible to transfer experiences via VR, we have some good evidence to think that these sorts of experiences are unlikely to change our behavior in the medium or long term. Why would that be?

AR and VR as Sympathy Machines

Earlier, I talked about Avenanti et al.'s (2010) research on bias and pain perception. Their study was one small piece of a much larger puzzle regarding bias and bias reduction tools. Ever since the introduction of the Implicit Association Test (IAT) by psychologist Anthony Greenwald in 1998, the study of bias and bias reduction has been dominated by the IAT approach.[19]

Calvin Lai, along with two dozen colleagues from universities around the world, conducted one of the largest analyses ever on the effectiveness of bias reduction techniques (Lai et al. 2016). They were interested in seeing to what extent bias reduction approaches changed measures of both explicit and implicit bias (including IAT score) over the short, medium, and long term. Lai et al. looked at nine different anti-bias strategies, which they grouped into the following categories: "Exposure to Counterstereotypical Exemplars, Appeals to Egalitarian Values, Evaluative Conditioning, and Intentional Strategies to Overcome Biases" (2016, 1004). Their results are sobering:

> Only five of the nine interventions replicated the immediate reduction effect ... and none had an effect after a delay. Implicit preferences rebounded quickly, possibly within several hours. We also found that the interventions did not have an effect on explicit preferences, implicit-explicit relations, or support for affirmative action.
>
> (Lai et al. 2016)

In a larger follow up study, they successfully replicated these results:

> The most dramatic result of the current research is simultaneous strong evidence for short-term malleability in implicit preference and little evidence for long-term implicit preference change just a couple of days later ... Not only did all of the interventions tested here fail to reduce implicit racial preferences, but they also failed to reduce explicit racial prejudice. Moreover, the interventions did not change support for affirmative action or internal/external motivations to respond without prejudice.
>
> (Lai et al. 2016)

The implication of Lai et al.'s study is thus that current approaches to reducing bias don't work particularly well or last very long. While none of the interventions they studied used VR or AR for bias reduction, evidence on the effectiveness of these technologies for reducing bias is mixed (Martingano, Hererra, & Konrath 2021). As I'll suggest, one reason for this is simply due to there being so little data. There just aren't very many studies, at this time, that use VR as a tool to reduce implicit and explicit bias, and those that exist either examine only very short time periods or often focus narrowly on changes to IAT score.[20]

Another reason why this may be the case has to do with how empathy-enhancing simulations are designed. As we've already seen, simulations high in perspectival fidelity and context-realism are likely to be experienced differently than simulations with comparatively lower scores on these dimensions. For example, one study on VR and bias focused on Dutch citizens and measured the (virtual) distance that they kept between themselves and a simulated stranger waiting at a bus stop (Dotsch & Wigboldus 2008).

In addition to IAT score, Dotsch & Wigboldus used two additional measures of implicit bias. One was a measure of the distance subjects kept between themselves and the virtual stranger. The other was a measure of galvanic skin conductance (GSC). GSC is often associated with measures of arousal or excitement (high/low) and can be associated with many emotions, including fear. One advantage of using multiple measures is that we can draw more concrete conclusions about unconscious biased behavior from their data than we can by focusing on IAT scores alone. Their study was both perspectively faithful and context-real. Perhaps unsurprisingly, these researchers discovered that

> native Dutch participants on average kept more distance towards [sic] avatars with a Moroccan appearance compared to avatars with a White appearance. Also, participants showed a small but reliable increase in skin conductance levels when approaching a Moroccan avatar as compared to when approaching a White avatar.
>
> (Dotsch & Wigboldus 2008)

Well designed simulations really do seem to generate virtually real experiences and, in this case, show that virtually real prejudice tracks real-life prejudice even down to microaggressions like how much distance people keep from strangers. It's important to note that this study was not intended as an empathy-enhancing simulation. Subjects of the study were not told that they were embodying anyone else during their experience. I've argued that because VR and AR can't deliver on its promise to give us the experience of "what it's like" to be someone else, empathy-enhancing simulations are unlikely to be ethical to use. However, there may be a way to harness the power of well designed simulations for the purpose of bias-reduction. What's important is that we design these simulations to enhance *sympathy* and not empathy.

Some researchers at this point in the argument may argue that if empathy-enhancing VR simulations are successful enough at making us less biased (less racist, less ableist, less xenophobic, more concerned with homelessness, the environment, the welfare of non-human animals, etc) then we might be morally justified in developing and using them even despite the sorts of issues that I've been raising against them in this chapter. There's some evidence to think that these critics have a point.

Domna Banakou and her colleagues, for example, created a simulation where White (Spanish) subjects were exposed to between one and three ten-minute VR Tai Chi sessions led by a virtual teacher (Banakou et al. 2016). Banakou varied whether her subjects had a virtual Black or White body and assessed racial bias using the IAT. Banakou's results are interesting in the sense that subjects did appear to reduce their IAT scores over the course of a week.[21] Banakou has also argued that inhabiting other bodies will have similar kinds of results. For example, in a follow-up study, she had subjects inhabit the body of Albert Einstein and found that subjects reduced their age bias, again measured using the IAT (Banakou et al. 2018).

These are interesting results, and Banakou is not alone in reporting changes in her subjects after they go through an experience of virtually embodying other people. Similar results have been found when people inhabit virtual bodies of those older or younger than they are or when they embody stereotypically attractive avatars. Reporting on Ahn et al.'s cow study, Stanford psychologist Jeremy Bailenson claimed that subjects, at least in the short term, claimed to eat less meat after going through their cow experiences (Bailenson 2018). I've already argued that such results are likely to be brittle (i.e., that they're unlikely to survive once the truth is revealed and may even undermine trust in larger social systems). I think that we can raise three additional problems for results like these.

In this chapter, I've been looking at the ethics of virtual embodiment as a form of empathy enhancement. In that context, I've argued that it's unethical to design a simulation whose function is to nudge its users by making them believe that they now know what it's like to be someone or something that they aren't already in a base-case relation with. Insofar as VR or AR embodiment studies like Banakou's don't rely on the explicit (or implicit) promise of getting access to "what it's like" to be someone or something else, then they may, perhaps, be able to avoid these problems. It's an open question whether or not this is possible.[22] If it isn't possible, then these simulations fall prey to the arguments I've laid out above (they lack transparency and are deceptively manipulative as nudgers). Another potential problem with studies like these is their reliance on the IAT as a measure of biased behaviors. The IAT, by itself, may or may not correlate well with actual reductions in discriminatory actions given the individual variability of IAT scores over time (Hofmann & Schmitt 2008). Pairing IAT with other measures can help close this gap, as for example Dotsch and Wigboldus (2008) did in their bus stop study.

However, even if it were possible, there's another ethical problem with designing and using simulations like these. If subjects are not being given access to the first-personal "what it's like" experiences of others when they run through these simulations then, as we discussed in Chapter 2, they must be engaging in theory-theorizing about the experiences of other people. When we theory-theorize someone else's psychology, we rely on assumptions (stereotypes) about what that person is like and what it might be like to be them. This need not be too problematic when we know the person well but becomes more problematic the less we know the person as we rely on more and more general (and thus less and less accurate) assumptions about them.

What this amounts to is that subjects are guessing, using their own internalized stereotypes of race (or gender, or disability, etc.) to adjust their experiences while in VR or AR. Victoria Groom and her colleagues have argued, for example, that when we embody a virtual body different from our own, our perception is affected by the triggering of stereotypes associated with those whose bodies we're inhabiting:

> Participants embodied as Black demonstrated levels of implicit racial bias favoring Whites that were higher than participants embodied as White. In addition, the only difference in explicit racism was between

Whites and nonWhites ... The implicit and explicit results provide support for the stereotype activation theory, which predicted that participants embodied as Black would demonstrate the highest levels of implicit racial bias and levels of explicit bias comparable to participants embodied as White ... People aware of stereotypes express implicit bias when stereotypes are activated, regardless of their agreement with them, while people low in prejudice control the explicit expression of stereotypes.

(Groom, Bailenson, & Nass 2009)

In a world where stereotypes and privilege are internalized at a young age and where the triggering of stereotypical representations can have harmful effects on others, it's wrong to create simulations that encourage users to rely on this approach, even in the name of bias reduction. Just like the students who were playacting homelessness on the campus quad, subjects in studies like these aren't any closer at understanding what it's like to be someone else and we're all arguably harmed if we're encouraged by a simulation to rely on stereotyped thinking in these ways.[23]

Luckily, we can avoid all of the problems I've been raising for virtual embodiment by changing how we design our simulations away from empathically embodying others and towards sympathetically engaging with them instead.

Although Lai et al. (2016) found that interventions aimed at mitigating biases usually produced only (fairly weak) short term effects, they discuss a few intervention strategies that may prove more effective:

One area of promising evidence for long-term change is research involving prolonged everyday experiences. These interventions are primarily conducted outside of psychology laboratories. Prolonged interventions that have been successful in changing implicit preferences and stereotypes include: taking a semester long class on prejudice and intergroup conflict ... having an college roommate who is of a different race ... participating in a cultural music education program ... and taking a class with a female professor.

(Lai et al. 2016)

They also suggest that interventions aimed at younger populations may be more successful at producing longer term changes in implicit attitudes.[24] Interventions designed to work against our internalized stereotypes also seemed to have some efficacy, especially, as they say above, when combined with repeated exposure. We have evidence to think that repeated tasks in virtual reality can, via neuroplasticity, affect neural structure itself (Cameirão et al. 2010). We also have some neuroscientific evidence to suggest that shifting a simulation away from empathy and toward sympathy may help effect more lasting positive changes in internalized bias.

Feng Sheng and colleagues (2014), for example, studied how Chinese subjects judged the pain of Chinese and Caucasian faces when they were asked to focus on different elements of these faces. In some cases, Sheng asked subjects to make judgments about the racial characteristics of the faces while in others they were asked to attend to whether the faces showed signs of being in pain. Although in neutral tasks Sheng found the same kinds of in/out-group biases in pain perception that Avenanti et al. (2010) found, things changed when subjects were asked to attend to specific aspects of the photos. Using fMRI, Sheng et al. claimed to find

> evidence that the racial bias shown in the ACC [anterior cingulate] and AI [anterior insular] activity in response to others' suffering can be reduced significantly by task demands that emphasize attention to others' emotional states. Relative to the race judgment task that emphasized race identity, the pain judgment task instructed participants to focus on others' personal feeling and significantly increased the ACC and AI activity to pain expressions of racial out-group individuals.
>
> (Sheng et al. 2014)

These results are consistent with Alison Martingano's meta-analysis of VR empathy simulations (Martingano, Hererra, & Konrath 2021). VR isn't good at enhancing cognitive empathy, though they did argue that emotional empathy can be affected by simulated experience. Correl et al. (2002; 2016), Xiao et al. (2016), Lai et al. (2016), and Sheng et al. (2014) give us a road-map for how to move forward. These researchers give us some reason to think that our internalized concepts can exercise top-down effects that structure how we perceive the world (i.e., that structural intersectionality is true for at least some of our experiential percepts). They also show us that long-term engagement with diverse peoples, experiences that contain counter-stereotypical representations of diversity, and an emphasis on the emotional states of others can all work to counter the triggering of internalized biases. So what would designing VR for sympathy instead of empathy look like? We have some real-world examples that can help us improve on problematic empathy simulations.

Michael Godlman, for example, is the director of the US Holocaust Museum. The Holocaust Museum, among the many things it's charged with, works to educate its visitors on the horrors of the Holocaust, and has experimented with VR technology to help it do this. Unsurprisingly, they've run into the same kinds of problems I've been discussing throughout this chapter, and their responses to these problems also give us a model for how to move forward:

> Goldman ... has discussed two issues that have come from displaying VR in the Holocaust Museum. Either the visitor minimizes their own experiences, where they think they should not feel bad for themselves,

say, because a friend died of cancer, because a Holocaust victim experienced something worse. Or, the visitor over-empathizes with a Holocaust survivor, where they think they know how it feels to be in the Holocaust. To combat these two scenarios Goldman treats visitors as "engaged witnesses" where they recognize the trauma of others without taking that trauma upon themselves.

(Thatcher 2019)

The concept of being an "engaged witness" works well with VR and AR. Being an engaged witness isn't about experiencing life through someone else's eyes. Instead, you are called on to bring your own sensibilities to bear on a simulation and its events. This is the difference between "simulation" and "in-their-shoes" empathizing. While VR and AR aren't able to let us engage in "simulation" empathy, they are especially well suited to bringing *us* to simulated worlds and, if well designed (if they are geared to produce virtually real experiences) they can have us respond *as if* the events in virtual worlds are real. Designing a simulation for sympathy, instead of empathy, requires that we bring users to important scenes to experience it *for themselves*.

A colleague of mine in our Art and Art History department, Takeshi Moro, has been working to develop sympathy enhancing simulations of exactly this kind. Hotchkiss (2019) has developed simulations of Japanese internment camps built in the US during the Second World War. As a part of that project, Moro, using a 360-degree camera, interviewed Japanese-American citizens forced to live in these camps. Moro's simulations place viewers back into (digitally recreated versions of) these camps to hear from survivors and learn from their experiences. Viewers are not told that they'll be experiencing internment life first-hand, as that would mean they would be in an empathy simulation. Instead, they're engaged (sympathetic) witnesses to these events and to the people who actually had these experiences.

Simulations like those used at the US Holocaust Museum or those developed by Takeshi Moro are thus prime examples of sympathy-enhancing simulations. Building on this work, we can imagine sympathy-enhancing VR and AR simulations that direct viewers to counter-stereotypic representations of others, where they are tasked with focusing more on the emotions that their virtual partners are experiencing and less on their out-group markers. Such simulations are not only free of the morally problematic assumptions built into empathy-enhancing simulations, they're also much more likely to work by harnessing the actual capabilities of VR and AR technologies for "in-their-shoes" empathy.

In the two chapters that follow, I'll be looking at more ways in which VR can change us. In the next chapter, I'll sum up the arguments made thus far in the book into a general code of ethics for VR and AR, that also includes material from existing technological codes of ethics produced by tech-related professional organizations. In Chapter 7 I'll focus more specifically on augmented reality to draw out ethical issues that are underappreciated. Although I've argued that the line between AR and VR is at least somewhat artificial, AR is already being used not only to give us more information about the world we live in, but also will raise significant social/political questions. Because we're likely to use AR to change how we appear to others, and because AR gives users almost unlimited freedom in how they present themselves to others, we need a framework to prepare ourselves for questions about identity, representation, and corporate/political power that are likely to emerge from these transformations.

Notes

1 Military drone operators, for example, are said to experience a form of homuncular flexibility (and what we've been calling presence) as they work (Martin 2015). This feeling of telepresence coupled with their sense of homuncular flexibility may be one reason why drone operators experience post-traumatic stress at about the same rates as traditional US Air Force pilots (Otto & Webber 2013). The experiences of traditional Air Force pilots and drone operators may be similar enough, as a result of perspectival fidelity and context-realism, that a drone operator's experience is virtually real despite the mediated nature of their work.

2 As I've noted throughout the book, AR is at a somewhat earlier stage of development than VR and hence AR-based empathy-enhancing simulations are very rare. In Chapter 7 I will consider, at some length, the future of AR simulations like these and their effects on our sense of self, our connections to our bodies, and corporate responsibility.

3 I'll say more about nudges in this chapter but, in short, nudges can be thought of as any intentional manipulation of an environment (real or virtual) whose purpose is to guide (but not determine) users toward making specific choices (Sunstein 2015).

4 Keeping in mind that the context-realism and perspectival fidelity of a simulation help us to determine whether it's well-designed for in-their-shoes empathizing. Simulations high in both, when experienced by someone with the right mix of psychological traits, are likely to produce virtually real experiences and thus generate ecologically valid simulations and realistic behavior. It's not impossible, in other words, for a well-designed simulation to really help us see how we might respond to the trolley problem or ticking-time-bomb cases.

5 In light of what I've said earlier, it's certainly better to have subjects walk on all fours than to walk as most humans do but, given the vast differences in cognition and embodiment between humans (of all kinds and abilities) and cows, no amount of haptic feedback or locomotion will allow VR designers to bridge this gulf. Philosophers may already see a connection between the assumptions of Ahn's simulation and well-known philosophical arguments like Thomas Nagel's "What is it like to be a bat?" (1974). I'll say more about these arguments later in the chapter.

6 I use the term intersectional here but as we'll see in this section, "structural" intersectionality is separable from claims about systems of oppression. One can be committed to structural intersectionality about experience without simultaneously being committed to intersectional theories of political oppression. Readers who might chaff at this framing of intersectionality are invited to view structural intersectionality as a claim about the top-down effects (in the

psychological sense) of internalized concepts on the content of individual experience (Collins & Olson 2014). While there's an active debate about the degree to which top-down effects exist (Firestone & Scholl 2015), I'll sidestep that debate here.

7 Emphasis added.

8 Psychologists and philosophers of perception sometimes distinguish between "top-down" and "bottom-up" influences on perception. Top-down influences occur when conceptual categories affect how we see the world (e.g., imagine the difference between what a botanist sees when she looks at the rainforest compared to what a cattle rancher might see). Bottom-up influences refer to the influence of a stimulus and our nervous system activity on perception. These are non-conscious, though as we'll see, at least some of these processes appear to be influenced by our internalized concepts as well.

9 He recognizes, for example, that a congenitally blind person will have a very different experience of color concepts than a sighted person even though, as is obvious, they're both human and both have color concepts (Nagel 1974, 440).

10 We could imagine, for example, that an enterprising VR or AR designer could add a voice-over track to *Preterna* that would fill in these details (that they had been trying to become pregnant, are an adult, can easily afford to raise the child, and so on) but such an approach would succumb to the same background/foreground problems raised by Goldie (2011) in Chapter 2 and which made simulations of the Bridge variant of the trolley problem impossible.

11 Avenanti's (2010) subjects were Italian and North African, which might affect how we understand the racial contexts of their study. As we'll see, similar results about pain perception translate to Chinese contexts as well.

12 Emphasis added.

13 As I'll argue in this section, dishonest nudges, because of their deception, are always unethical because they subvert user autonomy. Assuming that we're dealing with users whose autonomy is deserving of respect (i.e., that we're not dealing with children or those judged to lack competence), deception harms such users *even if* they are being deceived for genuinely paternalistic reasons.

14 Cass Sunstein argues that a person's welfare is promoted when they're "better off, *as judged by themselves*" (Sunstein 2015, 429).

15 For example, we might imagine that inflating the calorie information for unhealthy foods and minimizing the calorie information for healthier foods might work as an even more powerful nudge than simply giving people honest information about calories. Such a nudge is unethically manipulative because it is deceptive in at least two ways. First, it straightforwardly gives users false information. Second, it misrepresents itself as an educative nudge (it relies on the fact that users will assume that the information is accurate) and thus it lacks transparency. In fact, the lack of transparency is necessary in order to make a nudge like this work! Deception in a nudge can render it an unethical tool for manipulation even if manipulation is not always unethical.

16 Arguably, such a nudge is not transparent at all since the information it represents is false and, were that known, the nudge would cease to be (as) effective. I'll argue that something very similar is happening with respect to the ethics of empathy simulations.

17 Harris' view is of course not shared by all philosophers of technology. My point in raising his objections here is to show that even in a best-case scenario, VR empathy enhancement can't avoid more basic questions related to moral enhancement technologies more generally. Because this chapter is focused on the ethical issues unique to VR technology, I'm largely sidestepping these other concerns.

18 Cass Sunstein (2015), in discussing the roles of transparency and autonomy for the ethics of nudges, sounds a similar note as Harris: "An organizing idea is that when one is being manipulated, one is being treated as a kind of 'puppet on a string.' No one wants to be someone's puppet, and it is especially bad to be a puppet of the government" (443).

19 The IAT is not without its critics, and other models of measuring implicit bias are often used instead of or in conjunction with the IAT. As we'll see when we look at the meta-analytic data regarding bias and bias prevention, it's wise for us to consider broadening our conception of implicit bias beyond the IAT.

20 Lai et al. (2016) note, for example, that it's still an open question whether changes to one's IAT score correlate well with changings in biased *behavior*. If we want to change the baised ways that people treat others, then we'll need to make sure that we're not designing interventions aimed at reducing one's IAT score while keeping biased behavior unchanged.

21 Banakou's results in this study are especially confusing. She finds that subjects' IAT scores decreased most heavily after two exposures (a third exposure where White subjects inhabit a Black body actually seemed to *increase* bias relative to the second exposure). It's not clear what, if anything, was producing such a strange effect. If virtual embodiment in a Black body by White subjects lowered IAT scores, as Banakou seemed to suggest, we would expect more embodiment to lead to more significant decreases. As I'll argue below, one possible interpretation of these results is that practice at the IAT test itself may have lowered the scores and that demand characteristics may be responsible for bringing them back up during the third exposure. Such an interpretation is consistent with Lai et al. 2016.

22 It's hard, for example, to avoid the problem of demand characteristics when subjects inhabit a virtual body that is clearly not coded to represent their own racial identity. It would take some clever experimental design to avoid this confounding variable.

23 Recall that in Joshua Correl's research (Correl et al. 2002; 2016) on the police officer's dilemma, reliance on ethnic and racial stereotypes was primarily responsible for how subjects made sense of what someone might be holding (gun or cellphone?). Unsurprisingly, but depressingly, subjects were much more likely to disambiguate objects in the hands of a Black man as a weapon. In a surprising follow-up study using differently colored (light/dark) robots, a similar reliance on color stereotypes guided subjects' decisions about what dark colored robots were holding (Bartneck et al. 2018). Because stereotypes are, by nature, inaccurate heuristics for making sense of the world around us, their use becomes problematic when it has material effects (the welfare criterion) on the lives of others.

24 When it comes to VR and AR, the ethics of exposing children, especially those under 13, to simulations is fraught. I'll say more about this in upcoming chapters.

References

Ahn, S.J., Bostick, J., Ogle, E., Nowak, K., McGillicuddy, K., & Bailenson, J.N. 2016. Experiencing nature: Embodying animals in immersive virtual environments increases inclusion of nature in self and involvement with nature. *Journal of Computer-Mediated Communication*, 21 (6), 399–419.

Autismity. 2016. Retrieved from https://theautismsimulator.com/

Avenanti, A., Sirigu A., & Aglioti, S.M. 2010. Racial bias reduces empathic sensorimotor resonance with other-race pain. *Current Biology*, 20 (11), 1018–1022.

Bailenson, J. 2018. *Experience on demand: What virtual reality is, how it works, and what it can do.* London: Norton.

Banakou, D., Hanumanthu, P.D., & Slater, M. 2016. Virtual embodiment of white people in a Black virtual body leads to a sustained reduction in their implicit racial bias. *Frontiers in Human Neuroscience*, 10, 601. doi:10.3389/fnhum.2016.00601.

Banakou, D., Kishore, S., & Slater, M. 2018. Virtually being Einstein results in an improvement in cognitive task performance and a decrease in age bias. *Frontiers in Psychology*, 9, 917. doi:10.3389/fpsyg.2018.00917.

Bartneck, C., Yogeeswaran, K., Ser, Q.M., Woodward, G., Sparrow, R., Wang, S., & Eyssel, F. 2018. Robots and racism. In *Proceedings of the ACM/IEEE international conference on Human–Robot Interaction.* Chicago: ACM/IEEE, 196–204.

Bernstein, S. 2020. The metaphysics of intersectionality. *Philosophical Studies*, 177 (2), 321–335.

Blumenthal-Barby, J.S. 2013. Choice architecture: Improving choice while preserving liberty? In Coons, C. & Weber, M. (Eds.), *Paternalism.* Cambridge: Cambridge University Press.

Bovens, L. 2009. The ethics of nudge. In Grüne-Yanoff, T. & Hansson, S.O. (Eds.), *Preference change: Approaches from philosophy, economics and psychology.* Dordrecht: Springer, 207–219.

Cameirão, M.S., Bermúdez i Badia, S., Oller, E.D., & Vershure, P.F.M.J. 2010. Neurorehabilitation using the virtual reality based Rehabilitation Gaming System: Methodology, design, psychometrics, usability, and validation. *Journal of Neuroengineering and Rehabilitation*, 7, 48.

Carastathis, A. 2014. The concept of intersectionality in feminist theory. *Philosophy Compass*, 9 (5), 304–314.

Chalmers, D. 2017. The virtual and the real. *Disputatio*, 9 (46), 309–352.

Cogburn, C., Bailenson, J., Ogle, E., Tobin, A., & Nichols, T. 2018. *1,000 cut journey.* ACM SIGGRAPH 2018, Virtual, augmented, and mixed reality, Article no. 1. Vancouver, British Columbia, August 12–16.

Collins, J., & Olson, I. 2014. Knowledge is power: How conceptual knowledge transforms visual cognition. *Psychonomic Bulletin and Review*, 21, 843–860.

Combahee. 2015. The Combahee River collective statement. United States [web archive]. Retrieved from the Library of Congress, https://www.loc.gov/item/lcwaN0028151/

Correll, J., Park, B., Judd, C.M., & Wittenbrink, B. 2002. The police officer's dilemma: Using ethnicity to disambiguate potentially threatening individuals. *Journal of Personality and Social Psychology*, 83 (6), 1314–1329.

Correll, J., Cloutier, J., & Mellinger, C. 2016. Discriminating perception. *Psychological Inquiry*, 27 (4), 281–285.

Couch, C. 2016, April 29. Disability-simulating VR promotes empathy. *MIT Technology Review*. https://www.technologyreview.com/2016/04/29/160444/disability-simulating-vr-promotes-empathy/

Crenshaw, K. 1989. Demarginalizing the intersection of race and sex: A black feminist critique of antidiscrimination doctrine, feminist theory, and antiracist policies. *University of Chicago Legal Forum*, 140, 139–167.

Crenshaw, K. 1991. Mapping the margins: Intersectionality, identity politics, and violence against women of color. *Stanford Law Review*, 43 (6), 1241–1299.

Cummings, J., & Bailenson, J. 2016. How immersive is enough? A meta-analysis of the effect of immersive technology on user presence. *Media Psychology*, 19 (2), 272–309.

Dotsch, R., & Wigboldus, D.H.J. 2008. Virtual prejudice. *Journal of Experimental Social Psychology*, 44, 1194–1198.

Firestone, C., & Scholl, B.J. 2015. Can you experience "top-down" effects on perception? The case of race categories and perceived lightness. *Psychonomic Bulletin and Review*, 22, 694–700. https://doi.org/10.3758/s13423-014-0711-5

Gasdaglis, K., & Madva, A. 2019. Intersectionality as a regulative ideal. *Ergo: An Open Access Journal of Philosophy*, 6 (44), 1287–1330. https://doi.org/10.3998/ergo.12405314.0006.044

Goldie, P. 2011. Anti-empathy. In Coplan, A. & Goldie, P. (Eds.), *Empathy: Philosophical and psychological perspectives*. Oxford: Oxford University Press, 318–330.

Greenwald, A.G., McGhee, D.E., & Schwartz, J.L.K. 1998. Measuring individual differences in implicit cognition: The Implicit Association Test. *Journal of Personality and Social Psychology*, 74 (6), 1464–1480. doi:10.1037/0022-3514.74.6.1464.

Grimsley-Vaz, E. 2018. Creator of "1000 Cut Journey" uses VR to help white liberals understand racism. Moguldom.com. Retrieved from https://moguldom.com/152786/creator-of-1000-cut-journey-uses-vr-to-help-white-liberals-understand-racism/

Groom, V., Bailenson, J.N., & Nass, C. 2009. The influence of racial embodiment on racial bias in immersive virtual environments. *Social Influence*, 4 (1), 1–18.

Ham, D. 2018. *I am a man*. Retrieved from http://iamamanvr.logicgrip.com/

Hansen, P.G., & Jespersen, A.M. 2013. Nudge and the manipulation of choice. *European Journal of Risk Regulation*, 3, 3–28.

Harris, J. 2013. "Ethics is for bad guys!" Putting the "moral" into moral enhancement. *Bioethics*, 27 (3), 169–173.

Hofmann, W., & Schmitt, M. 2008. Advances and challenges in the indirect measurement of individual differences at age 10 of the implicit association test. *European Journal of Psychological Assessment*, 24 (4), 207–209.

Hotchkiss, S. 2019, June 18. In San Jose's Japantown, contemporary transience takes on historical weight. KQED. Retrieved from https://www.kqed.org/arts/13859833/transient-existence-artobjectgallery-san-jose-japantown

Huxley, A. (1932/2010). *Brave new world* (11th ed.). New York: Vintage.

Iñárritu, Alejandro G. 2017. Carne y arena (virtually present, physically invisible). Retrieved from https://www.lacma.org/art/exhibition/alejandro-g-inarritu-carne-y-arena-virtually-present-physically-invisible.

Kluwe, C. 2014, March. *How augmented reality will change sports … and build empathy*. Video. TED Conferences. https://www.ted.com/talks/chris_kluwe_how_augmented_reality_will_change_sports_and_build_empathy

Lai, C.K., Cooley, E., Devos, T., Xiao, J.Y., Simon, S., Joy-Gaba, J.A., & Nosek, B. A. 2016. Reducing implicit racial preferences: II. Intervention effectiveness across time. *Journal of Experimental Psychology: General*, 145 (8), 1001–1016.

Martin, M. 2015. *Remote-split operations and virtual presence: Why the Air Force uses officer pilots to fly RPAS*. International Symposium on Aviation Psychology, Wright State University, Dayton OH.

Martingano, A.J., Hererra, F., & Konrath, S. 2021. Virtual reality improves emotional but not cognitive empathy: A meta-analysis. *Technology, Mind, and Behavior*, 2 (1). https://doi.org/10.1037/tmb0000034

Nagel, T. 1974. What is it like to be a bat? *The Philosophical Review*, 83, 435–450.

Noggle, R. 2018. Manipulation, salience, and nudges. *Bioethics*, 32 (3), 164–170. doi:10.1111/bioe.12421.

Ogle, E., Asher, T., & Bailenson, J. 2018. Becoming homeless: A human experience. Virtual Human Interaction Laboratory. Retrieved from http://vhil.stanford.edu/becominghomeless/

Otto, J.L., & Webber, B.J. 2013. Mental health diagnosis and counseling among pilots of remotely piloted aircraft in the United States Air Force. *MSMR*, 20 (3), 3–8.

Parsons, T.D. 2015. Virtual reality for enhanced ecological validity and experimental control in the clinical, affective and social neurosciences. *Frontiers in Human Neuroscience*, 9 (650).

Parsons, T.D., & Rizzo, A.A. 2008. Affective outcomes of virtual reality exposure therapy for anxiety and specific phobias: A meta-analysis. *Journal of Behavior Therapy and Experimental Psychiatry*, 39 (3), 250–261.

Pinch, T. 2010. Comment on "Nudges and cultural variance." *Knowledge, Technology and Policy*, 23 (3–4), 487–490.

Ramirez, E. 2017. Empathy and the limits of thought experiments. *Metaphilosophy*, 48 (4), 504–526.

Ramirez, E. 2019. Ecological and ethical issues in virtual reality research: A call for increased scrutiny. *Philosophical Psychology*, 32 (2), 211–233.

Ramirez, E. 2020. How to (dis)solve the Gamer's Dilemma. *Ethical Theory and Moral Practice*, 23, 141–161.

Ramirez, E., & LaBarge, S. 2018. Real moral problems in the use of virtual reality. *Ethics and Information Technology*, 20 (4), 249–263.

Ramirez, E., & LaBarge, S. 2020. Ethical issues with simulating the Bridge problem in VR. *Science and Engineering Ethics*, 26 (6), 3313–3331. doi:10.1007/s11948-020-00267-5.

Ruckmann, J., Bodden, M., Jansen, A., Kircher, T., Dodel, R., & Rief, W. 2015. How pain empathy depends on ingroup/outgroup decisions: A functional magnet resonance imaging study. *Psychiatry Research: Neuroimaging*, 234 (1), 57–65.

Ruíz, E. 2017. Framing intersectionality. In Taylor, P.C., Alcoff, L.M., & Anderson, L. (Eds.), *The Routledge companion to philosophy of race*. New York: Routledge, 335–348.

Schwartz, A. 2017, March 20. Confronting the "shocking" virtual-reality artwork at the Whitney Biennial. *The New Yorker*. Retrieved from https://www.newyorker.com/culture/cultural-comment/confronting-the-shocking-virtual-reality-artwork-at-the-whitney-biennial

Selinger, E., & Whyte, K.P. 2010. Competence and trust in choice architecture. *Knowledge, Technology and Policy*, 23 (3–4), 461–482.

Sheng, F., Liu, Q., Li, H., Fang, F., & Han, S. 2014. Task modulations of racial bias in neural responses to others' suffering. *NeuroImage*, 88, 263–270.

Sunstein, C. 2015. The ethics of nudging. *Yale Journal of Regulation*, 32 (2), 414–450.

Tannenbaum, D., Fox, C.R., & Rogers, T. 2017. On the misplaced politics of behavioral policy interventions. *Nature. Human Behavior*, 1 (7) 0130.

Thapan, P. 2016, November 15. Finally! You can be pregnant in virtual reality. *Vice*. Retrieved from https://www.vice.com/en/article/3d53kk/you-can-be-pregnant-in-virtual-reality-finally

Thatcher, S. 2019. VR and the role it plays in museums. Retrieved from: https://ad-hoc-museum-collective.github.io/GWU-museum-digital-practice-2019/essays/essay-9/

Won, A.S., Bailenson, J., & Lanier, J. 2015. Homuncular flexibility: The human ability to inhabit nonhuman avatars. *Emerging Trends in the Social and Behavioral Sciences: An Interdisciplinary, Searchable, and Linkable Resource.* doi:10.1002/9781118900772. etrds0165.

Xiao, J.Y., Copping, G., & Van Bavel, J.J. 2016. Perceiving the world through group-colored glasses: A perceptual model of intergroup relations. *Psychological Inquiry*, 27 (4), 255–274.

6 Putting It All Together

A Code of Ethics for VR/AR

Over the course of this book, I've argued for interconnected ideas about virtual and augmented reality technologies and how design choices can radically change how we experience simulated content. Sometimes, this has been with an eye toward keeping us honest (e.g., by arguing that VR and AR can't do many of the things developers often claim they can); at other times I've been more interested in arguing how to use these technologies to improve ourselves (or, at least, to avoid harming ourselves).

In this chapter, I bring all of these ideas together to offer a unified code of ethics for these technologies. My intention here is to speak not only to professional philosophers but also to those actively developing VR and AR content (those who are not *also* philosophers!). The goal is to distill the collected insights of the prior chapters into a set of principles that can be consulted throughout all stages of development for a VR or AR product (or service).

To do this, I'll begin by surveying the ethical landscape for AR and VR developers as it exists today (i.e., the early 2020s) in order to argue that much of what makes VR and AR technologies uniquely compelling is also responsible for ethical pitfalls that have yet to be fully appreciated by commercial developers and researchers alike. I'll borrow heavily throughout this chapter from an ethical toolkit developed by Shannon Vallor, Brian Green, and Irina Raicu for the Markkula Center for Applied Ethics (An ethical toolkit n.d.) and intended to help designers integrate ethical thinking into the development of their technological products or services. Those familiar with value sensitive design (Friedman & Hendry 2019) will find elements of the Markkula Center's approach familiar. The key feature of these approaches is that they provide a justification (and process) for including values, especially ethical values, into all aspects of the design and development process for technological products. While these approaches attempt to provide guidelines for applying ethical values to the design process, I'll also take a closer look at the actual codes of ethics for the two largest tech-oriented professional organizations in the world today: the Institute of Electrical and Electronics Engineers (IEEE) and the Association for Computing Machinery (ACM).

DOI: 10.4324/9781003042228-6

Drawing from sources like these will help us pull together my own work on the elements of virtually real experience into a unified code of ethics for VR and AR. The rest of the chapter will then focus on what I think are the most compelling use cases for the code: developing AR and VR content for children, unethical experimentation on human subjects, the ethics of the "Gamer's Dilemma" and simulated wrongdoing, and the pre-emptive steps we need to take as VR and AR become integrated into the basic structures of society.

The Ethical Landscape Today

The two largest professional organizations for those in technology-related fields are the Institute of Electrical and Electronics Engineers (IEEE) and the Association for Computing Machinery (ACM).[1] Together, these organizations have well over 500,000 members spread across the globe. They, and their partner organizations, hold hundreds of conferences every year bringing academic and professional members in contact to advance research, share experiences, and to help protect the global reputation of, and the public's confidence in, technologists.

Perhaps the most obvious way in which these organizations do this is through the development and enforcement of codes of ethics. These codes are intended to make clear what it means to do ethical work and to be a virtuous engineer or software developer, and set both aspirational and practical goals for good work practices. The ACM, for example, states that their "[c]ode is designed to inspire and guide the ethical conduct of all computing professionals, including current and aspiring practitioners, instructors, students, influencers, and anyone who uses computing technology in an impactful way" (ACM 2018).

What I'll argue in this chapter is that these codes make a good, but often imperfect, starting point for thinking about the ethics of emerging technologies. These codes are, of course, malleable and so they're designed to change over time as new technologies are developed and introduced into our world. VR and AR, I've been arguing throughout this book, are revolutionary technologies that introduce distinctive new ethical issues and, as such, both organizations should adapt their codes to better address these considerations. The sheer scope of these organizations means they play an influential, and global, role in shaping the ethics of VR and AR development. To begin, let's look at some of the key elements of the IEEE's (2020) code of ethics:

- To hold paramount the *safety, health, and welfare of the public*, to strive to comply with ethical design and sustainable development practices, to protect the privacy of others, and to disclose promptly factors that might endanger the public or the environment
- To improve the *understanding* by individuals and society of the *capabilities and societal implications* of conventional and emerging technologies, including intelligent systems

- To seek, accept, and offer *honest criticism of technical work*, to acknowledge and correct errors, to be honest and realistic in stating claims or estimates based on available data, and to credit properly the contributions of others
- To *treat all persons fairly and with respect*, and to not engage in discrimination based on characteristics such as race, religion, gender, disability, age, national origin, sexual orientation, gender identity, or gender expression
- To *avoid injuring others*, their property, reputation, or employment by false or malicious actions, rumors or any other verbal or physical abuses[2]

The claims made throughout this book can be seen as an extended analysis of how VR and AR technologies can affect the safety and welfare of individuals and the public (e.g., by intentionally or unintentionally generating harmful virtually real experiences in users), and to show how, when, and why, VR and AR technologies can sometimes fail to treat persons fairly and with respect (e.g., via deceptively manipulating empathy-enhancing nudges). As a statement of values, the IEEE code contains a powerful set of claims about the importance of understanding technological effects on welfare, safety, and individual fairness. Encouraging technologists to be accepting of criticism is also important. Sometimes, despite our best intentions, our products can have issues we did not foresee and we need to be open to changing (even canceling) development when such criticisms arise.[3] Beyond that, however, the IEEE code of ethics is fairly threadbare and it largely ignores the role of systemic impacts that technologies can have which are harder to trace back onto individual harms (e.g., algorithmic bias).

The ACM spends much more time developing and explaining their code of ethics. As an organization, they have split their code into four sections, each of which outlines the values that ground that section of the code, along with the professional expectations that those values create for its members. The code then spends a considerable amount of time explaining specific cases in which the code applies. In that sense, the ACM code is more than a statement of values. It attempts, as I will also attempt in this chapter, to tie specific professional obligations back to a shared set of values and to explain how that connection operates. At its core, the ACM and IEEE codes share commitments to harm-reduction, respect for privacy, and designing with the common good in mind.

The ACM takes this a step further by articulating the consequences of accepting these values to the actual work of designing and implementing new applications or services. For example, in connection with the basic shared values of avoiding harm and designing for the common good, the ACM code then shows how these values generate special duties on developers when their systems are poised to become integrated into the infrastructure of society:

3.7 Recognize and take special care of systems that become integrated into the infrastructure of society. Even the simplest computer systems have the potential to impact all aspects of society when integrated with everyday activities such as commerce, travel, government, healthcare, and education. When organizations and groups develop systems that become an important part of the infrastructure of society, their leaders have an added responsibility to be good stewards of these systems ... As the level of adoption changes, the ethical responsibilities of the organization or group are likely to change as well. Continual monitoring of how society is using a system will allow the organization or group to remain consistent with their ethical obligations outlined in the Code. When appropriate standards of care do not exist, computing professionals have a duty to ensure they are developed.

(ACM 2018)

It's important, and correct, for the ACM to place special emphasis on technologies that are poised to become integrated into societal infrastructures, if only for the consequentialist concerns that such technologies will impact more of us than a niche-driven technology. VR and AR are both technologies poised to become integrated into the infrastructure of society in much the same way that computers and telecommunications technologies became integrated in the 20th century, and as such they deserve the highest possible scrutiny now, before such an integration has taken place.

For example, in 2019, the Facebook Corporation (parent company of the Oculus Corporation) began publicly discussing their multimillion dollar project to create something they called *Horizon*. *Horizon*, as envisioned by Facebook, will be a VR social media space which they hope will become the dominant social media space of the 21st century (Kaser 2019). Whether Facebook's hopes for *Horizon* come to pass (as of this writing, *Horizon* is in an invite-only beta testing phase), it seems likely that virtual social spaces like *Horizon* or one of its competitors will become an increasingly dominant part of our lives throughout this century. In the next chapter, I'll consider the ethical landscapes of these more mature AR and VR worlds, including how we might use these spaces to help make us wiser, more empathic people. As a harbinger of things to come, in 2021 Facebook faced significant backlash from users of its Oculus VR hardware over the corporation's plans to force users to engage with VR advertisements in order to play one of their more popular VR games, *Blaston* (Walker 2021). Users inundated Steam, a software purchasing platform, with negative reviews of the game until the Facebook Corporation walked back their decision to force in-game VR advertising. It's for reasons like these that we need to explore more concrete applied approaches to AR and VR ethics. Although this may seem like a relatively minor problem, VR and AR technologies are already equipped to deliver much more intrusive advertising. For example, some current standalone VR hmds have inside-facing cameras that can track a user's gaze while they run applications. Such cameras (currently used largely to optimize graphic card usage so that only areas where users are focusing receive high fidelity rendering) can be used to deliver unavoidable ads.

When, and under what conditions, should corporations be permitted to deliver advertisements in the context of VR and AR simulations? An ethics for VR and AR can help us begin to address these questions.

Beyond professional organizations like the ACM and the IEEE, researchers working with VR and AR systems have sought to fill the current gap in VR/AR-specific codes with their own suggestions for how to think about the ethical use of these systems. I'll briefly touch upon two influential proposals put out by separate labs. Michael Madary and Thomas Metzinger (2016), researchers at Johannes Gutenberg-Universität Mainz, have offered what is perhaps the most extensive set of ethical principles for VR (and to a lesser extent, AR). For example, throughout their article Madary and Metzinger argue that:

- Extended interactions with VR environments may lead to more fundamental changes, not only on a user's psychology but also on a biological level
- The comprehensive character of VR simulation plus the potential for the global control of experiential content introduces opportunities for new and especially powerful forms of both mental and behavioral manipulation, especially when commercial, political, religious, or governmental interests are behind the creation and maintenance of the virtual worlds (e.g., the recent controversy over advertisements in *Blaston*)
- No experiment should be conducted using virtual reality with the foreseeable consequence that it will cause serious or lasting harm to a subject
- Scientists and the media need to be clear and honest with the public about scientific progress, especially in the area of using VR for medical treatment
- Torture in a virtual environment is still torture. The fact that one's suffering occurs while one is immersed in a virtual environment does not mitigate the suffering itself
- Avatar ownership and individuation will be an important issue for regulatory agencies to consider
- Following a code of ethics is not the same as being ethical. A domain-specific ethics code, however consistent, developed, and fine grained future versions of it may be, can never function as a substitute for ethical reasoning itself

Madary and Metzinger's proposals are admirable and largely, I think, correct. Indeed I see my proposals in this chapter as largely compatible with their efforts. For example, I share their concerns with prolonged use of VR and AR and the impacts that prolonged use will have on all sorts of human capacities. Shannon Vallor (2015), for example, has raised concerns about the possibility of technologies causing what she termed "moral deskilling." Moral deskilling, in Vallor's terms, might follow from the use of technologies designed to make decisions or discriminations once thought to require human input.[4] One could easily imagine a world where VR or AR systems process morally sensitive information on our behalf in such a way as to lead to moral deskilling, and we should be extremely skeptical of turning over morally sensitive decisions over to automated processes.[5]

Where all three proposals fall short is with offering specific guidelines for how researchers and commercial developers ought to think about their workflows to produce ethical VR and AR applications. For example, the IEEE, ACM, and Madary and Metzinger codes all correctly ask developers to avoid causing unnecessary harm to users. Similarly, Madary and Metzinger rightfully claim that following a code of ethics is not the same thing as acting ethically. I'll touch on the nature of technomoral virtues in just a moment. The sorts of values enshrined in these codes of ethics are good starting points for us but on their own are not sufficient.

Why would we think this? First, considerations to avoid harm, to replace moral judgments with algorithmic decision making, and so on are not unique problems for VR or AR applications of technology. Presumably we should always try to avoid causing more harm than necessary when we act (whether or not we're engineers, computer scientists, philosophers, or presidents). We should always aim to be honest about the risks inherent in our work and we should not oversell our abilities or the advantages of our products regardless of what abilities or products we're talking about.

Second, Madary and Metzinger, for example, sometimes do go into details about the unique ways in which VR and AR can impact these values (e.g., VR is more likely to raise ethical issues due to its potential to cause depersonalization and derealization), but a more useful set of guidelines would move beyond the general concerns about "avoiding harm," about being concerned for the public good, or with being honest about the limitations of the technology. The central question remains: how is it that we're to enact these principles given the unique features of VR and AR technologies? How do we become good (virtuous) developers, researchers, and users of VR/AR content?

One advantage to the approach I'm advocating for is that we can think about the ethics of harm avoidance using the specific design tools implied by the concepts of perspectival fidelity and context-realism. These concepts give us much more fine-grained advice for how we *ought* to design simulations to avoid harm, about what sorts of design parameters can be adjusted to help us avoid making a simulation that moves from being (mere) violent entertainment to being an instance of "virtual torture."[6] Later in this chapter, I'll have more to say specifically on the ethics of virtual actions and how they fall out of the code I'm proposing. Before I get there, I want to first say more about what it means to think about a code of ethics in the context of a character-based approach to the ethics of technology.

One positive aspect of the three codes I've examined is that they're friendly, but don't *require* us to adhere to many specific moral theories or frameworks in order to get on board with them. The same is true of value-sensitive design approaches along with the Markkula Center's ethics toolkit. We can value honesty and harm avoidance for their own sake as intrinsic goods, or because they tend to increase overall welfare, or because we believe that we would choose such values under fair starting conditions, or because we think they would benefit us directly. Indeed, one advantage of these codes is that they are also compatible with what we might call a "virtue" or "character" based approach to moral decision making and development. Such approaches have enjoyed a significant resurgence among ethicists over the last 30 years, and character-based approaches to the ethics of technology are increasingly common in the field (Vallor 2011; Vallor 2016; Harris 2008; Curzer 2018).

Many modern Western approaches to ethics, like Utilitarianism, Contractualism, Ethical Egoism, and Deontology, see their goal as deriving a set of ethical principles or rules meant to guide right action, the formation of laws, and so on.[7] The character-based approach to ethics takes instead as its starting point a question over what sorts of people we ought to be. In this sense this approach, like the ACM code, accepts that acting in line with some moral code or set of rules doesn't equate to moral action. This is, at least in part, because they view moral situations as inherently complex. Such situations must be taken on their own terms. A rule like "avoid causing unnecessary harm" is unhelpful because it doesn't tell us how to resolve specific cases. In order to apply the rule we would need to have already developed special qualities, talents, ways of seeing, and so on (what philosophers would call virtues) so that when we are confronted by these situations we are already trained to see what harms are relevant to the situation and then to balance other ethical variables that may be at play in any given context.

Ethics, on these views, can't be abstracted into a set of rules or decision procedures (or algorithms) to tell us how to decide cases. Instead, these ethical frameworks see as their chief goal the development of the sorts of capabilities (virtues) that good people need in order to be the sort of people who can make the right decisions on a case by case basis. Theorists like Vallor (2016) argue that certain technologies may make it more difficult for us to develop specific virtues and may themselves demand that we modify our current understandings of virtue in order to make sense of how we ought to work with, and in some cases incorporate, these technologies into our conception of the good life. Call these sorts of traits technomoral virtues. Ethical codes like the ones discussed so far are good moral principles but they don't go very far into helping us see not only what technomoral virtues are needed to develop ethical VR and AR content but also which technomoral virtues may be at risk of deskilling us as a result of these technologies. What new forms of seeing, what unique ethical variables, will we need to develop in order to answer questions about the ethics of creating virtual worlds or about augmenting our own reality? In what follows, I'll begin this practical turn by examining the "Ethics Toolkit" developed by the Markkula Center for Applied Ethics (Vallor, Green, & Raicu 2018).[8]

My intention here is to use the elements of the Markkula Center's ethics toolkit as a practical starting point for researchers and developers of VR and AR content, to provide concrete ways of incorporating ethical thinking into their work at various stages. The toolkit itself is not developed exclusively for VR or AR applications and the sorts of techniques it suggests incorporating would be equally at home in the 20th century as they would be in the 21st. Following this introduction to the toolkit, I'll connect the elements to a proposed code of ethics for VR that takes these elements into account. The toolkit is broken down into seven tools or procedures that can be used to minimize harm and develop technomoral virtues.

- Tool 1: Ethical risk sweeping: Ethical risk sweeping involves the sensible practice of anticipating the possible ethical (as opposed to merely legal) risks posed by a potential application or service. Risk sweeping can be improved by increasing the diversity of a the group doing the sweeping (see "Expanding the ethical circle")
- Tool 2: Ethical pre-mortems and post-mortems: This tool focuses on identifying systemic ethical failures of a project either before (pre-mortem) or after (post-mortem) it has failed. What ethical problems could (or did) go wrong to cause this application to fail?
- Tool 3: Expanding the ethical circle: Tech workers tend to be a relatively homogeneous group in terms of class, education, race, and gender (Rooney & Khorram 2020). Although this may happen for any number of reasons, one drawback to homogeneity is that the values, interests, and perspectives of those outside these groups are likely to be ignored or misrepresented. Expanding the ethical circle, as a strategy, asks groups to be more sensitive to this fact and to be more inclusive (along multiple dimensions) when selecting the members of a project or group
- Tool 4: Case-based analysis: it's a truism to say that similar cases should be treated similarly. Case-based analysis asks a group to consider whether there are any products or services similar enough to the one the group is considering developing such that they can learn from their mistakes and successes
- Tool 5: Remembering the ethical benefits of creative work: Ethical design and engineering isn't just about identifying risks and avoiding disaster; it's about a positive outcome: human flourishing, including that of future generations, and the promotion of healthy and sustainable life on this planet
- Tool 6: Think about the terrible people: This tool implores designers to change how they think about their user base. Instead of thinking about their ideal user, thinking about the terrible people, as a strategy, requires that developers imagine how someone who seeks to abuse or misuse their product might use it (e.g., content reporting and moderation functions may be designed with a virtuous user base in mind who will only report truly problematic posts, but designers should think about how those who might seek to abuse or brigade such a system can be prevented from doing so)
- Tool 7: Closing the loop: Ethical feedback and iteration: "Ethical design and engineering is never a finished task – it is a loop that we must ensure gets closed, to enable ethical iteration and improvement"

These tools are useful heuristics for how we might move beyond a compliance framework for ethical design, as the ACM Code also suggests, and toward more rigorously incorporating ethics into project workflows. Incorporating these tools into a unified code of ethics for VR and AR applications will yield a code that is not merely a statement of values but which also includes practical, action-guiding, recommendations for safeguarding ourselves from repeating some of the technological mistakes of the recent past. I'll here introduce and then later elaborate upon the code:

Table 6.1 A code of ethics for VR/AR

- Design simulations so that they are only as perspectivally faithful and context-real as necessary (i.e., leverage the dimensionality of perspectival fidelity and context-realism to fine-tune a simulation's likelihood of producing virtually real experiences to the minimum degree necessary to achieve your desired effect)
- The Equivalence Principle should guide design and implementation of VR and AR applications. The more likely a simulation is to induce virtually real experiences, the higher the justification needed to develop and use it
- VR and AR applications intended for children 13 years or older should receive additional scrutiny. Non-therapeutic VR and AR applications for children younger than 13 years of age should not be developed at this time
- Design VR and AR applications so as to avoid encouraging prolonged use (which may cause dissociation and/or derealization and thus may affect whether subjects experience simulated events as virtually real). This can include simple prompts ("You have been in VR/AR for 2 hours, perhaps it's time for a break?") or more restrictive means (i.e., shutting down the application for a set time)
- Applications intended to change a user's behavior need to be transparent, avoid manipulation, and further both the user and public good (this is doubly important for educative nudges which rely on truthful content for their nudging power). Do not develop "empathy-enhancing" simulations
- Content in VR and AR should be treated more conservatively than content in traditional media. It may be necessary to provide more restrictive content ratings for VR and AR simulations than screenbound content ratings for the very same content, given the possibility of these applications inducing VREs
- User data, including eye-tracking and biometric information, should not be collected unless necessary for application functionality. Users should have the right to determine what data is collected and how it is used. Additionally, unavoidable advertisements making use of biometric user data (like gaze tracking) should be avoided. Users should always be afforded the option not to attend to advertisements
- Any VR or AR application, including social VR/AR, that is likely to be integrated into the infrastructure of society must receive the highest level of ethical scrutiny[9]
- VR and AR applications should avoid including elements that encourage or incentivize moral deskilling and should, if possible, encourage the development of technomoral virtues

Source: Adapted from Ramirez, E., Tan, J., Elliott. M., Gandhi, M., & Petronio, L. 2021. An ethical code for commercial VR/AR applications. N. Shaghaghi et al. (Eds.). *INTETAIN 2020*, *LNICST 377*, 1–10.

One of the cornerstones of this code of ethics is what in Chapter 4 I called The Equivalence Principle. That principle, recall, is a way of framing the ethics of virtual experiences:

> If it would be wrong to allow a person to have an experience of something in the real world, then it would be wrong to allow a person to a *virtually real* analogue of that experience. As a simulation's likelihood of inducing virtually real experiences in its subject increases, so too should the justification for the use of the simulation.
> (Bliznyuk 2019; Ramirez 2018; Ffiske 2020; Ramirez & LaBarge 2018)

Throughout the book I've argued that virtually real experiences are one of the most important, and least well understood, features of VR and AR simulations. Any code of ethics for VR and AR applications must be sensitive to these experiences. I've argued that virtually real experiences appear to be governed by three interrelated features: perspectival fidelity, context-realism, and user psychology.

Although simulations can be designed so as to mitigate the negative impacts of virtually real experiences by carefully tailoring a simulation's relative degree of perspectival fidelity and context-realism, by attenuating content ratings for VR and AR content, and providing warning labels, accounting for user psychology is more difficult. As I've noted earlier, some users (those especially high on certain Big Five personality traits and those who may be more prone to dissociation and derealization) are more likely to have virtually real experiences than others.

Building ethical virtual worlds requires a coordinated effort between developers, content regulators, the public, and governmental officials. To see how the code would work at the practical level, I'll investigate a few of the elements of the code here that raise what are, in my judgment, issues that, because they're fairly unique to VR and AR content, are especially pressing.

Children as a Special Case

The owner's manual for Oculus Rift hmds cautions users that

> [t]his product is not a toy and should not be used by children under the age of 13, as the headset is not sized for children and improper sizing can lead to discomfort or adverse health effects, and younger children are in a critical period in visual development ... Adults should monitor children age 13 and older who are using or have used the headset for any of the symptoms described in these health and safety warnings ... and should limit the time children spend using the headset and ensure they take breaks during use. Prolonged use should be avoided, as this could negatively impact hand-eye coordination, balance, and multi-tasking ability. Adults should monitor children closely during and after use of the headset for any decrease in these abilities.
>
> (Oculus Health and Safety Manual n.d.)

This is a refreshing approach to ethical risk sweeping and of expanding the ethical circle to include children. Children are in many ways a unique user base for any product, in part because, as Oculus notes, they are "in a critical period" of development. Facebook's/Oculus' concerns about VR use negatively impacting a child's skills of multi-tasking, balance, and hand-eye coordination are important; however, other risks abound for children not covered by the Oculus Health and Safety Manual. This is especially true for very young children who are progressing through several critical phases of development. Children between the ages of 3 and 5, for example, are in the process of developing neurotypical derealization abilities (Li, Liu, Woolley, & Zhang 2019). Very young children, in other words, are still developing the abilities connected to understanding the difference between real events and fantasy events. It is normal at these ages for children to genuinely believe that pretend events are really happening, and also for them to develop away from these beliefs.

These development stages are important for at least two reasons. First, as noted in earlier chapters, our capacities for dissociation and derealization appear to play an important role in influencing whether or not someone is likely to experience simulated (or imaginary) content as virtually real. These capacities lie on a spectrum. The more likely we are to dissociate from experiences (or to lose track of the sense of our experiences as happening *to* us), the more likely we are to experience simulated content as virtually real. This likelihood will also be affected by a simulation's degree of perspectival fidelity and context-realism.

For this reason, childhood use of VR and AR content, especially by young children and especially when using perspectivally faithful and context-real simulated content, may make it more likely that such simulations may negatively impact normal development of the psychological capacities for neurotypical dissociation and derealization. Because this means that such children would then be more likely to be accidentally harmed by VR and AR content (among other harms that may happen to them in their real-space lives), it's especially important for developers of VR and AR content to avoid designing applications for children. Similarly, research on young children using VR and AR should be subject to even more IRB scrutiny than research on children is already subject to. Second, we already have some evidence that VR content is able to induce dissociation from real-life events in otherwise neurotypical adults (Aardema, O'Connor, Côté, & Taillon 2010).[10]

We might imagine someone here objecting to these claims for two different sorts of reasons. First, someone might claim that these arguments sound suspiciously like familiar older arguments that have been raised against children using any new form of media including books, radio, television, computers, and the internet (Segall 2009). On the one hand, I think that there is something to these arguments. After all, data on the effects of VR and AR use on children are scant and there is some merit in avoiding a moral panic over new technologies. However, despite the force of this objection, I think it's important for us to always advance cautiously with new technologies, especially when it comes to children. It may be, in the end, that children are not negatively impacted at all by VR and AR use and, if it turns out that VR and AR use doesn't have the negative impacts on childhood development that early studies suggest it does, then the precautions I'm proposing in this section may not be necessary after all.

On the other hand, we might imagine a different kind of interlocutor who argues in the opposite way. This critic may, instead, argue that VR and AR use does in fact affect neurotypical development in exactly the ways described, but then go on to argue that the deskilling that children would experience as a result of prolonged AR and VR use is actually not really harmful. In the same way, such a critic might say, as children who grow up using GPS technology may deskill their capacity for spatial memory (Dahmani & Bohbot 2020), prolonged VR and AR use may impact their capacities for dissociation, derealization, and so on. In both cases, the critic would argue, such deskilling would be compensated for by new upskilling of other capacities necessary to function in their new (virtual) contexts. Here too I think that our critic might be onto something. In response, I think it best for us to approach such technologies cautiously. We know that hand-eye coordination, balance, multi-tasking ability, and capacities for neurotypical dissociation and derealization are important to us here and now. While it may be possible that such capacities become less relevant, even irrelevant, in a world of ubiquitous AR and VR, caution is advised. This is especially true when we're speaking of children and their cognitive and physical development, as such alterations are very difficult, sometimes impossible, to change.

As we gain a better understanding of how prolonged use of VR and AR content affects our capacities to understand the difference between real and simulated content, we may have to make ethical decisions about whether the benefits of the virtual worlds that we build are worth the deskilling that such exposure may be causing, and whether the costs are worth whatever upskilling (or other benefits) they bring to children.

VR and AR Content

When discussing VR and AR game design in Chapter 4, I quoted Scott Stephan, director of games at FoxNext VR Studio. Stephan was noting that in their studio they had to learn the hard way to change their approach to game design when creating experiences in VR compared to screenbound media. While that may not be surprising (VR and AR hardware give us more options for how we move around spaces than are available for traditional PC and game consoles), Stephan's concern was less about how to design virtual *spaces* (level design, haptics, locomotion, etc.) and more about how users *experience the same content* differently in VR than in screenbound media: "scary experiences, horror experiences need to be really finely calibrated. If you see a horror movie on a screen, you have the abstraction.

It's not so frightening, and you know you're there for fun …" Because VR can so much more easily induce a sense of presence, Stephan cautioned that "no creature should be larger than the size of a small dog. Anything above that and you get this primal, lizard-brain thing of, 'Oh, this isn't a fun scare. It's a survival scare'" (On immersive virtual reality 2018).

There's a general lesson here for developers of VR and AR content: content that might be experienced as innocuous in traditional media may become problematic when experienced as a VR or AR simulation. It's a mistake, in other words, for developers to assume that they can port a game designed for screenbound media over to VR or AR without problems associated with how the content will be experienced. Stephan's call for finely calibrating horror experiences in VR is exactly the right recommendation to make and, in addition to Stephan's solution (i.e., altering the content), perspectival fidelity and context-realism are the vehicles that VR and AR designers can use to enact these calibrations without altering the content but by altering how it's experienced by users.

How might developers do this? We can imagine ways of, for example, controlling elements of perspectival fidelity like whether or not a simulation has a diegetic soundtrack, or meta-informational overlays to help developers fine-tune user experiences to avoid "survival scares" (which can be harmful to users and unprofitable to developers) and encourage "fun scares." As a general principle, TEP would recommend that content in VR and AR should be approached more conservatively than content in screenbound media.

To take a real example of this phenomenon (and one where developers accidentally generated controversy), let's look at some trouble that developer Rockstar North found themselves in when they designed *Grand Theft Auto 5*. The Grand Theft Auto franchise has often intentionally courted controversy. This is a franchise famous for allowing (even encouraging) players to murder sex workers, drive recklessly, and rewarding players for murdering NPCs (even innocent NPCs). Such criticism tends to come "from the outside" (from politicians and other moral or religious leaders) and not from the game's players so it's especially interesting that this particular controversy came "from within" in that respect. Rockstar North somehow managed to outrage their own players (people who, we can assume, not only know what sort of game they're playing but who enjoy playing games in the franchise).

Reports of this controversy generated a fair bit of negative publicity for Rockstar North:

> Dubbed "the most disturbing scene" in the game by the popular gaming site Eurogamer, the episode comes midway through a mission called "By The Book," which has players oscillate between two of the game's player characters – Trevor and Michael – as they work to hunt down a terrorist … Trevor chooses between using a variety of torture-friendly appliances like a wrench and electrified clamps to persuade the suspect

into divulging more information ... In one particularly ugly moment, the player makes a circular motion with the gamepad's joystick to wrench a tooth out of the suspect's mouth.

(Lejacq, 2013)

Rockstar North, in this case, should have engaged in more ethical risk sweeping than they did when they designed the "By The Book" mission.[11] Although *Grand Theft Auto 5* was not designed as a VR game, it is now possible to play it in VR as a result of user-made mods. Despite the fact that the game was not made for VR (or AR), the concepts of perspectival fidelity and context-realism can help us make sense of why the "By The Book" mission caused so much controversy and it can serve as an important lesson for what developers of VR and AR content need to think about as they design their own worlds.

Readers familiar with the Grand Theft Auto franchise know that its gameplay isn't designed with perspectival fidelity or context-realism in mind. Most gameplay takes place using a third-person perspective, non-diegetic information and soundtracks are common, NPC behavior falls far short of producing a convincing illusion of reasons-responsiveness, items emerge from the corpses of the dead, and players themselves are immortal within the gameworld.[12] In short, players have become accustomed, by virtue of these design features, to treat their experiences as unreal. Whether intentionally or not, almost all of these things were changed for the "By The Book" mission. Elsewhere, I've written that

> [t]he alterations made to the design of the "By The Book" make it more likely that this mission will generate virtually real experiences in players compared with other missions. It is *these features*, I claim, that made the scene so controversial for players. Player reactions to this mission also explain why so many virtual killings are not especially controversial: they don't generate virtually real experiences. It isn't, in other words, that subjects see all virtual killings as morally unproblematic. Subjects *will* see virtually real murders as problematic, as the "By The Book" mission demonstrates ... when a simulation's design is perspectivally faithful and context-real. They will tend to react to these virtual actions *as if* they were real, at least in the moment.
>
> (Ramirez 2020)

Developers hoping to avoid similar player discomfort (and the subsequent loss of potential revenue such discomfort may cause) should take heed, especially those developing simulations and games for VR and AR applications. Perspectival fidelity and context-realism are important parameters over which developers exercise complete control and which give them a lot of power over how users will react to the virtual worlds they build.

There is one further motivation for developers to take perspectival fidelity and context-realism into account when they design their game worlds:

should many more instances of "By The Book" controversies materialize, they may draw the attention of legislators who may impose far stricter regulatory requirements on simulation development than might otherwise be needed. As we spend more and more of our time in simulated spaces and in VR and AR gameworlds, regulation may become inevitable.

Nudges

Nudges, I noted in the last chapter, are intentional alterations to the "choice architecture" of a space meant to influence (but not force) a person's choices in that space (Sunstein 2015). Everything from a speed bump in the road designed to slow you down to a supermarket layout designed to get you to wander through the aisles counts as a nudge. In that chapter, I focused on the specific class of VR nudges that sell themselves as "empathy enhancers" and I argued that empathy-enhancing simulations are almost always unethical to design or use. This is because those simulations run counter to a commitment to structural intersectionality (the view that our concepts, especially our concepts about identity, can give structure to our experiences). If we're committed to structural intersectionality, something I argued was fairly common among philosophers, psychologists, and critical theorists, then what it means for *you* to experience something is much more complicated than what VR and AR simulations can represent.

While some experiences may genuinely be base cases (i.e., instances in which the differences between the user and the person they're empathizing with don't matter) most will not be. Empathy-enhancing VR and AR simulations promise their users that it's possible for very different people to literally share one another's experiences. I argued in Chapter 5 that such simulations not only misrepresent the nature of empathy but that they also run into an Intersectional Dilemma. Either structural intersectionality is true, and hence empathy-enhancing simulations are impossible, or it is false and they are not. Both options come with steep costs. While rejecting structural intersectionality would remove most of the problems I introduced for VR and AR empathy enhancing simulations, it runs against a convincing accumulation of empirical and theoretical data. Since most developers of empathy-enhancing simulations are committed to structural intersectionality, the sacrifice is probably not worth making. Better to design simulations to induce *real* sympathy than deceive users using fake empathy and risk a brittle, potentially counterproductive, nudge.

VR and AR empathy-enhancing simulations run into other problems too. Tool 6 from the Markkula Center toolkit, "Think of the Terrible People," looms large when we consider assessments of empathy-enhancing nudges. Although most of us might respond with care and concern when we put ourselves through experiences like *1,000 Cut Journey* (Cogburn et al. 2018), it's important for VR and AR designers to think of the terrible people. Imagine, for example, what a committed White Nationalist may experience in the same simulation? If structural intersectionality is even partially true, the answer shouldn't surprise us (such a person may very well *approve* of the racism they experience in the context of the simulation). Better, I argued in Chapter 5, to avoid promising users that they'll understand "what it's like" to be someone else. VR and AR can be useful tools for anti-bias work but not because they can enhance our empathy in this way.

It's hopefully clearer why using VR and AR for "empathy enhancement" is a tempting, but nonetheless problematic, option for developers. Sympathy-enhancing simulations (those that place the user in the position of an engaged witness to the simulation's events) are likely to fare better as ethical alternatives. However, beyond empathy and sympathy, there are other ways of using these technologies to nudge users. For example, informational AR overlays can make for useful, and novel, educative nudging tools. Imagine having detailed nutritional information about all of the foods offered in a restaurant (or allergy information or sourcing information, etc. etc.). Imagine also the usefulness of having AR overlays while you're out hiking and that, thanks to your AR lenses, you are now also able to identify the plant and animal life around you (and thanks to its built-in GPS, you can even find your way to your destination no matter how far afield you wander).[13] Such nudges can encourage us to make more informed decisions about what we eat, where we travel, and how we spend our money.

More controversial nudgers, those that might begin to arguably violate autonomy for example, might be used to affect *how* we see others. We can imagine, for example, a nudge that uses AR overlays that work in much the same way that photo filters currently do in order to play into (or against) local beauty norms. Users could, for example, require that others see them as they prefer to be seen (e.g., blemish-free, with different skin tones, hair colors, even racialized markers). Early versions of this technology already exist (Bonetti, Warnaby, & Quinn 2018)! For example, cosmetic surgeons and makeup manufacturers are developing AR technologies to allow their customers to see "before and after" images of themselves.

The ethics of nudgers like these is complicated. On the one hand, "Snapchat Dysmorphia," a term used to describe people with body dysmorphic disorder who want to surgically alter themselves so that they look as they do when they use Snapchat filters on their photos, is a real phenomenon brought about (or made worse) by the use of photo-filters (Ramphul & Mejias 2018). We might imagine that AR technologies like the one I'm discussing here might (might!) reduce the incidences of BDD-induced surgical alteration by replacing it with real-time AR digital alteration. If so, that would, at least on the face of it, be a positive consequence of this sort of technology. Additionally, we traditionally allow people wide latitude over how they present themselves to others, up to and including surgical alteration. However, you aren't alone if nudgers like these make you uncomfortable. The problem is in trying to figure out whether this discomfort has an ethical grounding. A moral analysis of this technology requires that we look at a lot of variables.

Is this technology actually promoting a user's welfare? The public good? Does it sidestep what some might see as the real problem behind BDD-induced cosmetic surgery (i.e., harmful and gendered conceptions of beauty)? Additionally, how we imagine technology and its capacities matters. Can individual users decide how they would see others or would users have control over how they're seen by others? Will there be any limits on how users can self-represent or how others can choose to see them? Turning the question back to empathy enhancement, we might imagine a version of this AR filter that allows a person to be perceived as being a member of different racial groups (this could change by the hour or minute, depending on user preference). The social and philosophical backlash over so-called transracial identities suggests that, at least in American culture, at least some of our identity concepts (like race) are more rigid than others (like gender).[14]

Such a technology, were it to become integrated into the basic structure of society, would require that we engage in serious conversations about whether we need to alter how we think about concepts like identity, fidelity, discrimination, privacy, and so on. Imagine how we may have to restructure our concept of romantic love if our partners can radically change how they look, sound, and even feel from moment to moment or, equally interestingly, if *we* can control how they look, sound, and feel to us.[15] These concepts are important enough to our private and political lives that any rush to develop and commercialize AR technology like this before these issues are settled could be disastrous (or at the very least seriously confusing).

As the ACM cautions us, any technology that stands to be incorporated into the basic structures of society deserve the highest level of scrutiny (and what could be more basic than the identity categories available to us?). It's enough to say that the ethics of VR and AR nudging is multifaceted and that figuring out whether a particular nudge (e.g., VR empathy simulations, AR avatar shaping, etc.) is ethically acceptable requires that we dig into specifics not only of how that technology works (data privacy, deception, manipulation, etc.) but also into the larger social/political consequences of adopting that technology. These are the kinds of ethical considerations that ought to enter any assessment of technologies that are poised to become a part of the basic infrastructure of society. I'll focus the next, and final, chapter on the ethical issues associated with these uses of AR.

Experimentation

In Chapter 4, I argued that Institutional Review Boards (IRBs), which, in most universities and research institutes are responsible for reviewing and approving any research that makes use of human (or animal) subjects, must take into account virtually real experiences when they assess the risks involved in VR and AR experiments. There, I said that IRBs need to be especially sensitive not only to the psychological traits that subjects in an experiment may bring to it (e.g., Big Five personality traits, dissociation, derealization, etc.) but also that they should routinely determine whether a simulation is more virtually real than it needs to be in order to carry out the experiment successfully.

In this section, I look at studies attempting to recreate classic thought experiments using VR in order to show not only why perspectival fidelity and context-realism are important for ecological validity but also why it's important for IRBs, and researchers themselves, to think about virtual harms that extend beyond virtual trauma, and I'll tie these considerations back to the proposed code of ethics.

The classic version of the trolley problem asks us whether we would be willing to switch a trolley from one track to another during an emergency. Switching the trolley would result in the death of one person who is stuck on the track but it would save five others who are stuck on the original track (Foot 1978). There are many versions of the trolley problem, some of which I discussed earlier. For example, an equally famous version created by philosopher Judith Thomson has gained widespread attention from moral psychologists for several reasons.[16] In Thomson's version, you are asked to imagine that you are standing on a bridge overlooking some trolley tracks while a situation unfolds below you. In this version, you're asked to imagine that you

> can see a trolley hurtling down the track, out of control. You turn around to see where the trolley is headed, and there are five workmen on the track where it exits from under the footbridge. What to do? Being an expert on trolleys, you know of one certain way to stop an out-of-control trolley: Drop a really heavy weight in its path. But where to find one? It just so happens that standing next to you on the foot-bridge is a fat man, a really fat man. He is leaning over the railing, watching the trolley; all you have to do is to give him a little shove, and over the railing he will go, onto the track in the path of the trolley. Would it be permissible for you to do this?
>
> (Thomson 1985)

For reasons which I've elaborated throughout this book, I think that thought experiments like the trolley problem, including this "Bridge" version of it, don't work very well in terms of getting us a clear picture of what our real moral judgments would be in situations like the ones they propose. The grip of the Thought Experiment Paradigm is still strong. I've argued elsewhere that the Bridge version of the trolley problem is all but impossible to recreate in VR (Ramirez & LaBarge 2020).

Why would this version of the trolley problem pose such difficulties for simulation in VR or AR? Simulation design matters and, in this case, the Bridge version of the trolley problem contains elements that make it especially difficult (I've argued that it's essentially impossible) to recreate. Although VR and AR could be used to more faithfully recreate a context-real and perspectivally faithful simulation of a world with trolleys, bridges, and people stuck on tracks, and while we might also be able to simulate other aspects of that world (like the time pressure involved in the decision itself), at least one essential feature of the Bridge version of the trolley problem can't be simulated in this way.

Recall that in Thomson's version you are "an expert on trolleys" and, as a result of that expertise you, on your own, come to believe that there is "one certain way to stop an out-of-control trolley: Drop a really heavy weight in its path." It's crucial, on this version of the trolley problem, that subjects in the experiment have come up with the idea to push the person on the bridge onto the tracks *on their own* (for the same reason, it's important in the classical version of the trolley problem that subjects realize on their own that they have the option to switch the tracks). Details like this matter, experimentally and philosophically and any simulation aiming to recreate these thought experiments must get them right. Why should we think this?

One reason why details like this can matter is because they're likely to affect how subjects think about their own responsibility for their decisions. Recall our discussion in Chapter 4 of Stanley Milgram's experiments. Although in that chapter I focused on Milgram's classic experiment (where Milgram himself, wearing a lab coat, stood near subjects and politely but firmly instructed them to "please continue" shocking the learners), Milgram actually conducted several variations of this experiment and these are instructive. In one version of the experiment,

> four (apparent) subjects arrived for an experiment on the effects of "collective teaching and punishment" on memory and learning. Only one of the individuals was a true subject. A rigged draw produced the learner and three teachers. The script called for the other "subjects" (assistants) to withdraw [from the study] at the tenth (150 volt) and fourteenth levels (210 volt), respectively. This defiance of authority was an extremely powerful influence on subjects, with only 10% continuing to follow instructions to the 450 volt level.
>
> (Miller & Collins 2010)

Changing an element of an experiment can radically influence its results. Recall that in Milgram's classic experiment *two thirds* of participants delivered the 450 volt shock. In this revised version, where subjects saw others resist Milgram's authority, only one in ten delivered the final shock. In part this might be because subjects were able to offload some (or most or even all) of their perceived responsibility for killing the learner onto Milgram in the original experiment (e.g., because he was a doctor, was asking them to continue, etc.). In this modification of the experiment things are different. Subjects *knew* they could leave without consequence and thus if they chose to stay they might assume that they bore full (or at least more) responsibility for their decisions.

Something similar is happening, I argue, with simulations of Thomson's Bridge version of the trolley problem. It's crucial to Thomson's original experiment that subjects are the source of the idea to push the man onto the tracks. However, how can anyone design a simulation that reliably gets subjects to form the idea, on their own, not only that a heavy enough weight can stop the trolley but that pushing the man onto the tracks is a viable option? Kathryn Francis and her colleagues attempted to recreate just this situation (2017).[17] Their design choices help us see why it's so difficult to simulate this version of the trolley problem. Francis and colleagues had

> [v]erbal instructions played during the 3D scenario and specific instructions were given prior to the experimental task, explaining that this task involved a joystick but that participants would be given a choice about whether they wanted to interact with the virtual object or not ... . After 30 seconds, verbal instructions informed participants that a trolley car was approaching ("*Look behind you, a train is coming.*") After a further 25 seconds, a second verbal dialogue then followed ("*Hey I am too far away but if you want to save the people you could push the large person on to the tracks and derail the train. If you're going to push him, do it now, but it is your choice.*").
>
> (Francis et al. 2017)[18]

In this case, Francis et al. did not faithfully reproduce Thomson's Bridge version of the trolley problem. My colleague Scott LaBarge and I have written elsewhere that in this specific study

> [t]he non-diegetic character of the voice-over leaves it entirely up to subjects to interpret its relevance. Does it represent the voice of God? Are subjects receiving audio from a nearby engineer? Are they meant to be hallucinating? Arguably, any of these interpretations diminishes the subject's responsibility. For example, if the voice-over is interpreted as a nearby engineer, the decision to push the man onto the tracks can be partially attributed to the engineer; if attributed to God, the voice becomes arguably the determiner of moral content. If the voice-over is meant as a hallucination, this may cause the subject to raise questions about the nature of the simulation itself.
>
> (Ramirez & LaBarge 2020)

Milgram saw his subjects' behavior change radically as he varied the experimental protocol for his obedience experiments (demonstrating that the situations, despite their similarities, were meaningfully different from one another), demonstrating the importance of holding experimental features constant. It's important for the Bridge problem that subjects get the idea to push the man onto the tracks all on their own if we want to see whether the judgments subjects make in VR are comparable to real life judgments or to the pen-and-paper thought experiment studies.

Lastly, it's important to note that even *if* it were possible to create a perspectivally faithful and context-real simulation of thought experiments like these, ethical considerations might still rule them out. We can already imagine why subjects might find themselves traumatized by virtually real experiences of pushing someone to their death, but other harms exist here too. If simulations like these are indeed possible then it means that VR and AR could be used in a way that can shape our behavior to make us more likely to treat people's lives as expendable in situations like these. Regardless of whether you think the right answer to the Bridge thought experiment is to push or not to push, IRBs (and experimentalists themselves) should factor the possibility of changing their subjects' character in this way as a part of any ethical analysis of VR and AR moral dilemmas. Better risk-sweeping and closing the loop (seeking feedback before and after experiments and revising plans in light of criticism) are an important part of the ethics toolkit.

Tourism

Tourism is interesting because it is simultaneously a common activity and an ethically fraught one (Holden 2003; Lovelock & Lovelock 2013; Malone, McCabe, & Smith 2014; Tribe 2009). The history of tourism is subject to intense controversy. Did tourism, as we understand it today, originate with the European aristocratic tradition of the "Grand Tour" (Zuelow 2015) or is this an unhelpfully Eurocentric way of thinking about tourism (Towner 1995)? What role do virtual and augmented realities play within larger ethical and cultural conversations about tourism? Is there such a thing as ethical and unethical VR and AR tourism?

It's pretty easy to see why traditional tourism, the kind that requires you to physically travel from one place to another, is ethically fraught. Travel, regardless of how it's done, always leaves behind a carbon footprint. It takes energy (usually fossil fuel energy) to drive, fly, or cruise from one place to another. In a world already experiencing negative impacts from anthropogenic climate change, purely touristic travel requires justification. Tourists can also indelibly impact the places they visit in both positive and negative ways. Tourism generates tourist money and tourist money can help provide economic opportunities and sustain communities that might otherwise not be economically viable. It can also do the opposite (destroying local economies and ways of life in the pursuit of tourism). Similarly, tourists have expectations (often racist, sexist, colonial, or otherwise exoticizing expectations) about the places they visit and locals may find themselves playing into these expectations to keep tourists happy and spending (or, worse, seeing themselves suffer economically if they refuse to play along).

Some forms of tourism (e.g., sex tourism) are especially problematic in that they often support unethical practices like human trafficking, animal poaching, and so on (Brooks & Heaslip 2019).[19] The 2020–2021 Covid pandemic caused many countries to restrict tourist travel on a scale not seen in the modern era. Companies eager to tap into the tourism industry during this time experimented with VR and AR travel opportunities to make up for these restrictions (Chen 2020). Even outside of pandemic travel restrictions, VR and AR tourism has been said to have many potential benefits. Aside from the reduced carbon footprint of VR and AR travel, virtual tourism could help decongest the globe's most popular tourist destinations, for example, by allowing some people to visit places virtually instead of physically and on their own schedules. Because simulated environments can be delivered at scale, it's likely that a virtual visit to Paris (even one paired with real-life meals-in-a-box of authentic French cuisine) will be significantly less expensive (and thus available to more people) than a real trip.

It's apparent that, at least in some cases, most of us already find virtual tourism morally preferable to real-life tourism. To take one example, we can appreciate what has been done with the Lascaux caves in France's Dordogne valley. These caves were long ago declared a Unesco World Heritage site. The network of caves contain some of the earliest, and grandest, artworks in human history and are an irreplaceable and one-of-a-kind link to our ancient past. Because of the delicate nature of the artworks and the pigments used to paint the walls of the cave, access to the caves is strictly controlled and tourists are not allowed inside. As a compromise, full scale replicas of the cave and its paintings have been built to allow tourists to experience the artwork without damaging the originals. These simulated Lascaux caves have proven very popular with tourists.[20] In cases where locations may be too delicate (for preservation or ecological reasons) to handle tourist visits, simulated experiences may be our best, and most ethical, alternative.

VR and AR can also allow people to engage in impossible tourism. It's currently (circa 2021) impossible for any of us to tour the surface of Mars, though VR simulations that combine images from Martian rovers (along with collected audio samples of Martian wind) can be gathered to reproduce a perspectively faithful and, arguably, context-real Martian tourist experience without the physical risks of space travel or the risk of contaminating the Martian surface with terrestrial life. In 2017, the American National Aeronautics and Space Agency (NASA) made this sort of simulation, *Access Mars*, available to the public (Good 2017).

Other forms of impossible tourism can be made possible via AR and VR. In Chapter 4 when discussing pedagogical VR and AR simulations, I noted that a VR simulation of the Buddhas of Bamiyan has already been created for use in the classroom (Johnson 2018). These towering 6th-century statues were destroyed by Taliban forces in 2001 and so are literally impossible to see today (though the enormous alcoves in which they were carved are still present). The VR and AR simulations of the statues are the only way for any of us to now get a first-hand sense of what they looked like. VR and AR can also give people more immersive ways of engaging in historical tourism (e.g., getting an on-the-ground view of Mesopotamian Ur streetlife, games at the Roman Colosseum, or a peek into the Imperial Court of 10th-century Japan).

Developers of VR and AR tourism, *especially* historical tourism of this kind, must be mindful not only to expand the ethical circle and think of the terrible people, but also to keep in mind the insights gained from our discussion of empathy-enhancing VR and AR simulations. First, simulations cannot, and should not be billed as, providing tourists with a "what it's like" experience in historical simulations. The Intersectional Dilemma will always loom impossibly large for simulations like these. Second, developers must be mindful to expand the ethical circle if they aim to represent peoples and cultures so that they include diverse members of those peoples and cultures in the design process. Tourist depictions of the peoples and cultures being visited or sold as destinations have a history of problematic representation (Hammett 2014; Caton & Santos 2009; Mellinger 1994). It's important not to repeat these practices with the images used for AR and VR tourist experiences.

Conclusion

The code of ethics presented in this chapter is an attempt to collect the insights produced not only by the IEEE and the ACM but also the insights of VR and AR researchers like myself, Michael Madary and Thomas Metzinger. Developers of VR and AR content, Institutional Review Boards, and commercial rating organizations would all benefit from an appreciation of the way that perceptival fidelity, context-realism, and user psychology can combine in VR and AR simulations to produce experiences that require careful ethical consideration: virtually real experiences.

In the next, and final, chapter I move beyond the concepts of perspectival fidelity and context-realism to explore the ethical risks involved with widespread AR technology. Some of these risks are easy for us to see (AR overlays can bring us advertisements anywhere at any time) and some, I'll argue, appear innocuous at first (cosmetic AR overlays) but raise ethical questions we may not be prepared to answer.

Notes

1 Though the ACM's code of ethics is significantly more robust than that of the IEEE, both are highly influential and worth a closer look by anyone interested in ethics and technology. The ACM code can be found here: https://www.acm.org/code-of-ethics. The IEEE's code of ethics can be found here: https://www.ieee.org/about/corporate/governance/p7-8.html

2 Emphasis added.

3 One of the Markkula Center's tools, for example, encourages developers to "Think of the Terrible People" (An ethical toolkit n.d.). This tool cautions us that "we must not envision our work being used only by the wisest and best

people, in the wisest and best ways. In reality, technology is power, and there will always be those who wish to abuse that power. This tool helps design teams to manage the risks associated with technology abuse." In that sense, the toolkit and the IEEE align. We must accept honest criticism of our technologies even if such criticism undermines years of development work.

4   The recent past has given us several examples of this sort of moral deskilling. The computer science department at UT–Austin, for example, has recently come under fire for using a machine-learning algorithm to help sort their graduate student applications. Similarly, on December 18, 2020, medical staff at Stanford Medicine staged a demonstration against the algorithmic process hospital administrators used to determine which staff members would be first in line to receive the Covid-19 vaccine. Similarly, in 2020 Californians rejected a proposed amendment to their state constitution that would have ended the state's "cash bail" system that would have turned over bail decisions to an automated process. Such examples show us that those turning over moral decision-making to algorithmic processes stand to make themselves worse at making such decisions on their own.

5   Not all technological deskillings are harmful or wrong! As a result of industrialization and material science advancement, a proportionally smaller number of people are capable of, or even familiar with, skills that were once quite common, even vital (e.g., animal husbandry, flint knapping, archival research using the Dewey decimal system). In a real sense, most of us are deskilled relative to those in the past in this respect. We've picked up new skills in part as a response to these changes. Vallor's warning is against *moral* deskilling in particular. We must be cautious that as technologies make certain normative aspects of our lives easier (e.g., sensors on cars can help us avoid harmful collisions) that we don't degrade our normative competencies as a result of turning over moral decisions (graduate admission decisions, parole decisions, etc.) to algorithmic processes. See also John Harris (2013) for an argument that "moral enhancements" would actually undermine moral capacities via a similar process. Harris doesn't use the term "deskilling" but his concerns about eroding our ability to discern ethical variables and to make difficult (i.e., fallible) decisions are part of what makes us moral agents in the first place. It's not clear, and I here do not intend to offer practical advice, how we should distinguish between morally benign forms of moral deskilling (e.g., losing the skill to hunt down basic morally relevant information to search engines) and moral problematic forms (e. g., offloading the job of allocating scarce resources to an algorithm). What's important is that we're aware of deskilling as a potential threat and actively consider the possibility that new technologies can deskill us as technologies are developed.

6   To their credit, Madary and Metzinger (2016) are clear that their intention is "to provide a first set of ethical recommendations as a platform for future discussions." In that sense, their project was a tremendous success as they carefully carved out some possible ethical concerns that, while not unique to VR and AR, certainly become more pressing with respect to these technologies.

7   In most ways, this is far too short an explanation. While categorical imperatives (e.g., lying is *always* wrong) are easy enough to understand as the sorts of universalist ethical principles a virtue ethicist may seek to reject, other theories may be more flexible. Philosophers actively disagree, for example, about whether utilitarian/consequentialist approaches to ethics are themselves best understood as a form of moral particularism (Bales 1971; Dancy 2004). If this turns out to be the case then the distinction between the consequentialist and the character-based theorist will turn on more foundational issues about what the point of ethics is. Without delving too far into the weeds, I mean to capture the distinction here

between two basic normative questions: what should I do and how should I live? The first is focused directly on our actions while the second concerns itself with what sort of people we should be.

8 Although I was not personally involved in the creation of the toolkit, the Mark-kula Center for Applied Ethics is located on the Santa Clara University campus and I am affiliated with, and have received funding from, the center. The elements of the Toolkit offer approaches relatively common in the Science, Technology, and Society (STS) field and readers may know these elements by different names depending on their background.

9 I devote the next chapter to an extended analysis of how augmented reality applications are likely to be integrated into the basic structure of society and why such an integration (useful and necessary as it might be) will introduce several problems that *need* to be addressed in advance in order to avoid significant metaphysical, epistemological, moral, and political problems having to do with identity.

10 Perhaps most surprising is that the simulated content used by Aardema, O'Connor, Côté, and Taillon was not particularly high in context-realism or perspectival fidelity, thus suggesting that VR and AR content that might be seen as permissible under TEP may nonetheless require us to be careful about the effect of such content on user psychology. This is one reason why long-term use, even by neurotypical adults and even of virtually *unreal* content, should be limited, at least for the moment.

11 Beyond risk sweeping, other tools from the Toolkit should also have been considered, including expanding the ethical circle (to include more non-White stakeholders, individuals with PTSD, and so on), thinking about the terrible people (for obvious reasons) and, given the controversy it actually generated, we can hope that *Rockstar North* has conducted an ethical post-mortem analysis (though a pre-mortem analysis would have probably prevented this issue from arising in the first place).

12 "Dying" in the franchise results in the player's character re-spawning at the nearest hospital (regardless of how destructive their pseudo-death was). It is impossible to receive a "game over" screen. This makes the gameplay more engaging (and maybe even more fun) but in doing so it profoundly diminishes the context-realism of the game universe.

13 Here the risk of deskilling (of losing the ability to keep track of where you are either by reading a map or developing internal spatial memory) should at least count against using AR in this way (though probably not decisively, depending on how reliable the technology itself is).

14 See, for example, the various responses to people like Rachel Dolezal who claim a transracial identity (e.g., Overall 2004; Tuvel 2017; Fernandes Botts 2018; Sealey 2018). At stake in this conversation is the difficult question about how to think about race as a social construction and whether (or to what degree) our socially constructed concepts of race are like (and are not like) socially constructed concepts like sex, gender, and even of bodily integrity and autonomy. My intention here is not to show that transracial identities are socially possible options but instead to show that these concepts are currently very much in tension and that AR applications like the one I'm talking about here would only further add to this confusion in ways currently missed by debates about trans-racialism and the metaphysics of identity as it relates to our physical presentation.

15 We might think that the answer is actually obvious: stop using the technology and you'll then be with your "real" partner. Such responses, however, beg the question against those who would identify with their augmented reality selves more than their purely biological self (imagine, for example, how some people

currently feel about their avatars in persistent open world games like *Second Life*). Another reason this wouldn't work well as a response is that it makes the problem seem like a personal one instead of a social one. There are, to be sure, plenty of people all over the world who have yet to use the internet (even if we exclude those who use it only sporadically). The fact that it's possible to live in such a way doesn't negate the fact that, as a species, we have had to deal with the ways that the internet and social media have altered our social, political, and private lives.

16 One of the most important reasons is that pen-and-paper versions of Bridge seem to be treated very differently by most people than pen-and-paper versions of the classic trolley problem (Greene et al. 2001). This difference may or may not be a real feature of moral judgment (that is, it may be an artifact of the Thought Experiment Paradigm and not really tell us much about our real-life moral judgments). In any case, the Bridge variant of the trolley problem, and other thought experiments like it, have since received much more attention.

17 Francis has since produced many innovative alterations to her original experimental paradigm in an aim to make her simulations of Bridge more perspectivally faithful (improving haptics) and context-real. Although this is an improvement, it's not clear that the most critical hurdle (producing a simulation where subjects naturalistically decide, on their own, that pushing the man is a solution to the problem) is either possible or ethical.

18 Original emphasis.

19 This remains true regardless of one's thoughts on the ethics of consensual sex work.

20 Much as with AR simulations, literal recreations like this get perspectival fidelity and (in this case) context-realism very cheaply and thus it makes sense that tourists have virtually real experiences of being *in* the real caves while they explore the fake caves. Virtually real experiences, and their components, can help us make sense of this phenomena too.

References

Aardema, F., O'Connor, K., Côté, S., & Taillon, A. 2010. Virtual reality induces dissociation and lowers sense of presence in objective reality. *Cyberpsychology, Behavior, and Social Networking*, 13 (4), 429–435.

ACM (Association for Computing Machinery). 2018. *ACM Code of ethics and professional conduct*. Retrieved from: https://www.acm.org/code-of-ethics.

An ethical toolkit for engineering/design practice. (n.d.). Retrieved from https://www.scu.edu/ethics-in-technology-practice/ethical-toolkit/

Bales, R.E. 1971. Act-utilitarianism: Account of right-making characteristics or decision-making procedure? *American Philosophical Quarterly*, 8 (3), 257–265.

Bliznyuk, A. 2019, February 5. Ethical guidelines in virtual reality: Towards a code of conduct in research. Retrieved from https://www.rm.wi.tum.de/fileadmin/w00bjc/www/Leadership_Learning_Innovation/Dokumente/Ethical_Guidelines_for_VR_TUM_11022019.pdf

Bonetti, F., Warnaby, G., & Quinn, L. 2018. Augmented reality and virtual reality in physical online retailing: A review, synthesis and research agenda. In Bonetti, F., Warnaby, G., & Lee Quinn, M. (Eds.), *Augmented reality and virtual reality in physical and online retailing: A review, synthesis and research agenda*. Cham, Switzerland: Springer. https://doi.org/10.1007/978-3-319-64027-3_9

Brooks, A., & Heaslip, V. 2019. Sex trafficking and sex trouism in a globalised world. *Tourism Review*, 74 (5), 1104–1115.

Caton, K., & Santos, C.A. 2009. Images of the Other: Selling study abroad in a postcolonial world. *Journal of Travel Research*, 48 (2),191–204. doi:10.1177/0047287509332309.

Chen, A. 2020, April 20. Is virtual travel here to stay, even after the pandemic subsides? *National Geographic*. Retrieved from https://www.nationalgeographic.com/travel/article/can-virtual-reality-replace-real-tourism-during-pandemic-and-beyond

Cogburn, C., Bailenson, J., Ogle, E., Tobin, A., & Nichols, T. 2018. *1,000 cut journey*. ACM SIGGRAPH 2018, Virtual, augmented, and mixed reality, Article no. 1. Vancouver, British Columbia, August 12–16.

Curzer, H. 2018. Yesterday's virtue ethicists meet tomorrow's high tech: A critical response to Technology and the Virtues by Shannon Vallor. *Philosophy and Technology*, 31 (2), 283–292.

Dahmani, L., & Bohbot, V.D. 2020. Habitual use of GPS negatively impacts spatial memory during self-guided navigation. *Scientific Reports*, 10, 6310. doi:10.1038/s41598-020-62877-0.

Dancy, J. 2004. *Ethics without principles*. Oxford: Oxford University Press.

Fernandes Botts, T. 2018. In black and white: A hermeneutic argument against "transracialism." *Res Philosophica*, 95 (2), 303–329.

Ffiske, T. 2020. Ethics in virtual and augmented reality in the 2020s. *Virtual Perceptions*. Retrieved from https://www.virtualperceptions.com/ethics-vr-ar-2020/

Foot, P. 1978. The problem of abortion and the doctrine of double effect. In *Virtues and vices and other essays in moral philosophy*. Oxford: Clarendon Press.

Francis, K., Gummerum, M., Ganis, G., Howard, I.S., & Terbeck, S. 2017. Virtual morality in the helping professions: Simulated action and resilience. *British Journal of Psychology*, 109 (3), 442–465.

Friedman, B., & Hendry, D.G. 2019. *Value sensitive design*. Cambridge MA: MIT Press.

Good, A.C. 2017, October 19. Take a walk on Mars – in your own living room. NASA. Retrieved from https://www.nasa.gov/feature/jpl/take-a-walk-on-mars-in-your-own-living-room

Greene, J.D., Sommerville, R.B., Nystrom, L.E., Darley, J.M., & Cohen, J.D. 2001. An fMRI investigation of emotional engagement in moral judgment. *Science*, 293 (5537), 2105–2108.

Hammett, D. 2014. Tourism images and British media representations of South Africa. *Tourism Geographies*, 16 (2), 221–236.

Harris, C.E. 2008. The good engineer: Giving virtue its due in engineering ethics. *Science and Engineering Ethics*, 14 (2), 153–164.

Harris, J. 2013. "Ethics is for bad guys!" Putting the "moral" into moral enhancement. *Bioethics*, 27 (3), 169–173.

Holden, A. 2003. In need of new environmental ethics for tourism? *Annals of Tourism Research*, 30 (1), 94–108.

Institute of Electrical and Electronics Engineers. 2020. IEEE code of ethics. Retrieved from https://www.ieee.org/about/corporate/governance/p7-8.html

Johnson, C.D.L. 2018. Using virtual reality and 360-degree video in the religious studies classroom: An experiment. *Teaching Theology and Religion*, 21 (3), 228–241.

Kaser, R. 2019, September 25. Ready player what? Facebook announces a new VR social network. The Next Web. Retrieved from https://thenextweb.com/facebook/2019/09/25/ready-player-what-facebook-announces-a-new-vr-social-network/

Lejacq, Y. 2013, September 17. "Grand Theft Auto V" torture episode sparks controversy. NBC News. Accessed 08/10/2018. http://www.nbcnews.com/technology/grand-theft-auto-v-torture-episode-sparks-controversy-4B11185864

Li, H., Liu, T., Woolley, J., & Zhang, P. 2019. Reality status judgements of real and fantastical events in children's prefrontal cortex: An fNIRS study. *Frontiers in Human Neuroscience*, 13, 444. doi:10.3389/fnhum.2019.00444.

Lovelock, B., & Lovelock, K. 2013. *The ethics of tourism: Critical and applied perspectives*. Abingdon: Routledge.

Madary, M., & Metzinger, T.K. 2016. Real virtuality: A code of ethical conduct. Recommendations for good scientific practice and the consumers of VR-technology. *Frontiers in Robotics and AI*. https://doi.org/10.3389/frobt.2016.00003

Malone, S., McCabe, S., & Smith, A.P. 2014. The role of hedonism in ethical tourism. *Annals of Tourism Research*, 44, 241–254.

Mellinger, W.M. 1994. Toward a critical analysis of tourism representations. *Annals of Tourism Research*, 21 (4), 756–779.

Miller, A., & Collins, B.E. 2010. Perspectives on obedience to authority: The legacy of the Milgram experiments. *Journal of Social Issues*, 51 (3), 1–19.

Oculus Health and Safety Manual. (n.d.) Accessed 12/21/2020. Retrieved from http s://securecdn.oculus.com/sr/oculusrifts-warning-english

On immersive virtual reality technology and the ethics questions it raises. 2019, October 8. *VR Life*. Retrieved from https://www.vrlife.news/immersive-experien ces-virtual-reality-design-ethics/

Overall, C. 2004. Transsexualism and "transracialism." *Social Philosophy Today*, 20, 183–193.

Plaisance, P.L. 2013. Virtue ethics and digital "flourishing": An application of Philippa Foot to life online. *Journal of Mass Media Ethics*, 28 (2), 91–102.

Ramirez, E. 2018. Ecological and ethical issues in virtual reality research: A call for increased scrutiny. *Philosophical Psychology*, 32 (2), 211–233.

Ramirez, E. 2020. How to (dis)solve the Gamer's Dilemma. *Ethical Theory and Moral Practice*, 23, 141–161. https://doi.org/10.1007/s10677-019-10049-z

Ramirez, E., & LaBarge, S. 2018. Real moral problems in the use of virtual reality. *Ethics and Information Technology*. https://doi.org/10.1007/s10676-018-9473-5

Ramirez, E., & LaBarge, S. 2020. Ethical issues with simulating the Bridge Problem in VR. *Science and Engineering Ethics*, 26, 3313–3331. https://doi.org/10.1007/s11948-020-00267-5

Ramphul, K., & Mejias, S.G. 2018, March 3. Is "Snapchat Dysmorphia" a real issue? *Cureus*, 10 (3), e2263. doi:10.7759/cureus.2263.

Rooney, K., & Khorram, Y. 2020, June 12. Tech companies say they value diversity, but reports show little change in the last six years. CNBC. Retrieved from https://www.cnbc.com/2020/06/12/six-years-into-diversity-reports-big-tech-has-made-lit tle-progress.html

Sealey, K. 2018. Transracialism and White allyship. *Philosophy Today*, 62 (1), 21–29.

Segall, Eric. 2009. In the name of the children: Government regulation of indecency on the radio, television and the internet – let's stop the madness. *University of Louisville Law Review*. Georgia State University College of Law, Legal Studies Research Paper no. 2009–01.

Sunstein, C. 2015. The ethics of nudging. *Yale Journal of Regulation*, 32 (2), 414–450.

Thomson, J. 1985. The trolley problem. *Yale Law Journal*, 94 (6), 1395–1415.

Towner, J. 1995. What is tourism's history? *Tourism Management*, 16 (5), 339–343.

Tribe, J. (Ed.) 2009. *Philosophical issues in tourism*. Bristol, UK: Channel View Publications.

Tuvel, R. 2017. In defense of transracialism. *Hypatia*, 32 (2), 263–278.

Vallor, S. 2011. Knowing what to wish for: Human enhancement technology, dignity, and virtue. *Techné: Research in Philosophy and Technology*, 15 (2), 137–155.

Vallor, S. 2015. Moral deskilling and upskilling in a new machine age: Reflections on the ambiguous future of character. *Philosophy and Technology*, 28 (1), 107–124.

Vallor, S. 2016. *Technology and the virtues: A philosophical guide to a future worth wanting*. New York: Oxford University Press.

Vallor, S., Green, B., & Raicu, I. 2018. Ethics in technology practice. The Markkula Center for Applied Ethics at Santa Clara University. Retrieved from: https://www.scu.edu/ethics/

Walker, I. 2021, July 23.VR players revolt against Facebook's in-game advertising plans. *Kotaku*. Retrieved from https://kotaku.com/vr-players-revolt-against-facebook-s-in-game-advertisin-1847152609.

Zuelow, E. 2015. *A history of modern tourism*. London: Red Globe Press.

7 AR and the Future of Selves

This chapter was written during the summer of 2021. Many parts of the world are slowly beginning to turn a corner in their handling of the Covid pandemic. Vaccines are being developed and distributed and they seem to be mostly effective even against the variants that have evolved over the last 14 months. In many places, the most restrictive regulations introduced at the height of the pandemic are being cautiously lifted. For those able to get one, vaccinations are letting them slowly start to reassemble daily routines that look a little like they did before the pandemic. I'm meeting with a former student (and a current friend) who is now in the middle of her philosophy Ph.D. program at a university on the East Coast. We've agreed to meet at a local coffee shop with some nice outdoor seating. The situation is as refreshing as it is ordinary (maybe refreshing *because* it's so ordinary). I arrive first and secure a table and pass time on my laptop. Then a shape appears in my peripheral vision and I instantly recognize my friend as she walks up. We haven't seen each other for at least eighteen months so we exchange happy greetings and get down to the business of updating each other on our lives, talking shop about projects we're working on, and enjoying our coffees. After more than a year of pandemic-induced hermitude, outings like this are a welcome return to something resembling normalcy and they're a good exercise of social skills that have grown rusty during our online pandemic existence.

There's a lot of interesting neuroscience and social psychology behind everyday interactions like this. A lot needs to happen for us to recognize our friends (and ourselves) over time. Friends might change how they look (sometimes a little, sometimes a lot). Despite the fact that people might look and sound a little different each time we meet them, in most cases these changes don't faze us. Sometimes it takes a minute ("could that be … yes it is!") but we usually recognize even long missed friends and family right away. In part this is because we have inherited a suite of heuristics (cognitive shortcuts) from our evolutionary ancestors that make these things possible. Each of these heuristics is built to make social interactions, especially the process of recognizing ourselves and other people, easy and natural for us. I recognize my friend at the coffee shop almost instantly even though it's

DOI: 10.4324/9781003042228-7

been well over a year since I've seen her, and this is how these mechanisms are supposed to work.[1]

This chapter is about what happens to these interactions, and the psychological heuristics we've evolved to make them possible, in a world where augmented reality (AR) technologies have become much more fully integrated into the basic structure of society than they are today. Our pre-AR ways of being social, of having relationships, and even of paying our taxes are likely to need an overhaul in this AR future. But wait, you might (and should!) be asking yourself: how would AR disrupt everyday interactions like meeting an old friend at a coffee shop?

In an earlier chapter, I argued that augmented and virtual realities were not distinct in kind but instead that they represented different points along a spectrum of extended reality experiences. Although I'll say more about this, we can start to appreciate the issues that AR is poised to introduce. These issues target some of the most essential aspects of our lives and who we think we are. We can see how this might work by imagining that I'm having a similar meeting with a friend in 2040 in a world where wireless access to an AR meta-informational universe is as natural a part of our lives as having omnipresent access to the internet is to most of us today. In such a world, I'll not only have a much easier time getting around, thanks to omnipresent AR directional overlays, but I'll also be able to get nutritional information about my food immediately, receive weather information in real time so I can get ready for that cold front that's making its way to the area in the next hour, and most importantly, we'll all have the power to customize how we look in these shared AR spaces. The changes we can make to our AR bodies can be more varied than the options we have today to change our physical bodies (e.g., by wearing different clothes, styling our hair, even cosmetic surgeries).

In this chapter, I'll argue that an AR future like this will render our evolved heuristics for reidentifying others useless and that we'll need new tools to help us manage. AR allows users to radically alter the ways they look, move, sound, and with the right haptics even alter the way they feel. How will I recognize my friend in 2040 if she could, literally, look, sound, move, and talk in ways radically unlike how she does today? She may even choose not to look recognizably human (in this world, perhaps many users would prefer to look *very* different from occasion to occasion). How will she recognize me if I decide that I'd like to look very differently than I did during our meeting in 2021? I won't be offering any definitive answers to these questions in this chapter (I'll leave that very hard problem to others whom I hope take up this question!). In that respect, this chapter is by far the most speculative and conjectural in the book. My goal is to convince you that we're not yet ready to address the issues that omnipresent AR technology raises for ourselves and our way of life.

I'll begin by laying out the basics of where AR and VR sit within what I've been calling the "XR spectrum." Although I'll be focusing on AR (because it seems like the most natural way of developing these technologies and integrating them into real-world social interactions), similar problems can arise if, by 2040, we're spending as much time in virtual realities as we do in augmented or vanilla realities. I'll then survey some of the cognitive heuristics that neuroscientists and psychologists believe we've evolved in order to engage in social reidentification (both self recognition but also, and especially recognizing others). These heuristics, I'll argue, may no longer work well (and may, instead, lead to mass confusion) in a world of omnipresent AR embodiment. Why is this?

One of the most natural extensions of AR technology is to empower people to represent themselves as they want to be seen. I'll spend some time looking at the ethics of body modification to make the point that even without AR and VR technologies, rational autonomous adults (and in many cases even juveniles) are afforded (legally and morally) wide discretion in how they shape their bodies to match their preferred ways of being seen. The ethical principles that justify these freedoms, I'll argue, naturally extend to AR embodiment and body modification. If I'm right, then this combination of facts will force us to address two questions about the ethics of AR body modification. Because these issues are, to my mind, novel and without much literature behind them, the ideas I present here aren't intended to answer these questions. Instead, I'll lay out what I think are the most plausible sorts of answers and show why each answer, despite seeming plausible, would force us to change our conceptions of our selves, change our regulations regarding privacy, and in some cases even force us to rethink what it means to be in a relationship!

The first issue we'll need to confront is a question about control. Who should have control over how people appear to others in omnipresent AR spaces? Here I'll survey four equally plausible answers: (a) users should have control over how they appear to others, (b) users should have control over how others appear to them, (c) states should regulate how users appear to one another, or (d) corporations should control how users appear when using their applications. Although each answer is plausible, I'll argue that each also has significant costs. None of these options will leave our current norms about the self intact and each will require not only that we change our expectations about what it is our selves *are* but also that we change our conception of privacy and its value.

The second issue is related to the first. Regardless of who we think should have control over how users appear in AR spaces, users will need some way of reidentifying the people they interact with in these spaces. As I argued above, evolution has outfitted us with many ways of keeping track of our social interactions but all of these presume a minimal degree of stability in terms of embodiment and physical appearance. AR body modification undermines all of these assumptions and makes these heuristics useless (arguably worse than useless as they're apt to generate confusion). As a result, we'll need to develop new ways of keeping track of who we interact with because the people we interact with can look, feel, sound, and even move differently from interaction to interaction. Here too I'll survey at least three plausible-seeming answers about how to provide a concept of identity in AR that is both stable (i.e., that can be tracked consistently) and sticky (i.e., that can be fixed to a distinct individual). These include (a) tracking identity using biomarkers (fingerprints, DNA, etc.), (b) tracking identity via the use of non-fungible tokens (NFTs), and (c) tracking user identity using correlated patterns of user activities.[2] As with the first issue about user representation, I'll argue that all of these answers come with steep tradeoffs. In most cases we won't be able to get sticky and stable AR identities without sacrificing something else of value. I'll also argue that, while the second and third options may be viable alternatives to explore as we consider the moral and legal landscape of our AR future, the first option is best left behind. This alternative runs against a strong current in contemporary ethics that our physical bodies are not essential to our identity.[3]

I'll close the chapter by considering other ways in which AR is likely to change our future. Relationships, for example, are likely to change, or at least become more complicated, in a world where how we look, sound, and feel can change on a user's whim. Notice that how we solve the issue of representational control impacts this (if users control how they appear to others, our lovers' appearance may be shifting constantly, if users control how others appear to them our lovers may be envisioning themselves being with someone who looks very different than we do). Traditional relationship norms no longer seem apt to work in such a context. They, like the heuristics for reidentification, are grounded on assumptions about the relationship between the body, its appearance, and identity. We'll need a new ethics of relationships (and at least a reconceptualization of concepts like monogamy, fidelity, and so on) for a world of omnipresent AR.

Equally worrisome is the possibility that instead of massively expanding individual freedom and diversity, AR embodiment and social norms will work together to radically *diminish* human diversity. In a racist, misogynist, nationalist world, it may turn out that very few people will voluntarily choose to have their AR appearance deviate from expected standards for physical appearance, movement, or voice. Some already go to great lengths to change their appearance to conform to (arguably harmful) beauty norms and it's not clear (barring a moral revolution to our concepts of beauty, race, gender, and so on) that people in the AR future will be much different. How can we make traction on the ethics of this kind of embodiment and its push to conformity? In the world today, a world before robust AR embodiment, there are limits to what we can do to change how we look, feel, and sound. We must, as a result, accept a certain minimal connection between our bodies and how we represent ourselves to others. We must accept a degree of chance, in other words, that ties us to our particular bodies. AR embodiment, however, removes this chance and gives people near infinite choice. Will such power ultimately benefit or burden us? I'll revisit a classic argument by philosopher Michael Sandel that, I argue, speaks to these questions.

Before getting deeper into all of these questions, let's first turn our attention to distinguishing between augmented and virtual reality and the idea of an extended reality spectrum. It's because AR has an especially strong connection to the actual physical world that it seems most poised to create the sorts of problems I examine in this chapter, at least in the near term.

AR, VR, and the XR Spectrum

Augmented reality applications, as I outlined in the third chapter, are part of a spectrum of simulations. This "extended reality" (XR) spectrum can be defined minimally by augmented reality simulations on one end, what we'll call "mixed reality" simulations in its central regions, and virtual reality simulations on the other end. Figure 7.1 illustrates this idea.

Writing about the ethics of modifying AR spaces, philosopher Erica Neely (2019) notes that AR is best described as "the use of technology in order to project digital content onto our experience of the physical world. Moreover, the content projected depends on (and adapts to) properties of the surrounding space." It is, in other words, the part of the XR spectrum where the world itself is overlaid with simulated content but in which the unaugmented world still serves as its own model.[4] The degree to which a simulation overlays content onto the world (and how much of the user's world consists of simulated content *not* overlaid onto an augmented world) will determine its location within the XR spectrum.

At the minimalist end of the XR spectrum we can imagine AR palette-swapping applications. A palette swap, in this instance, would be any application that simply changes how the world looks to us. We can imagine an application, let's call it *NoirWorld*, that allows users to experience the world in a high contrast monochromatic way so that users can experience the world so that it mimics the appearance of classic noir detective films. While this application would apply an augmented reality overlay onto the world, it does so minimally via palette swapping. The way the world smells and feels and the objects that exist in the world would all otherwise stay the same.

Luis' AR haunting application from Chapter 3 is a prime example of an AR simulation that fits near (but not on) the minimalist end of the XR spectrum. This is because Luis' simulation not only palette swaps AR

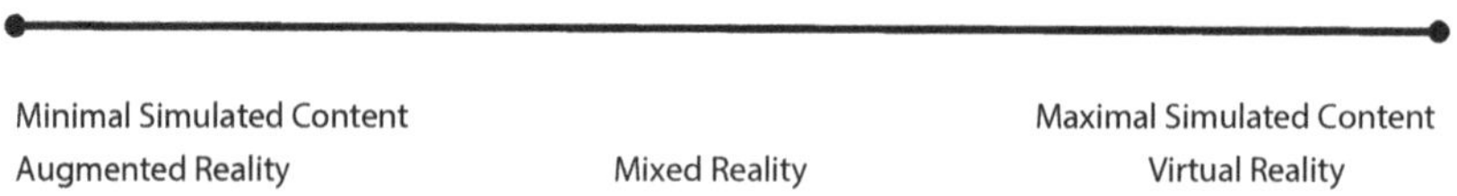

Minimal Simulated Content		Maximal Simulated Content
Augmented Reality	Mixed Reality	Virtual Reality

Figure 7.1 The XR spectrum

content onto an unaugmented world (e.g., the application can palette swap simulated dinginess onto his very-real walls to amp up the spookiness), it also overlays completely simulated content onto the world (e.g., causing ghostlike images at the periphery of his vision or in the reflections of mirrors), and it might also change the world in ways other than adding simulated content onto the world (e.g., the simulation may *remove* information in order to focus user attention or remove elements that diminish spookiness).

At the maximalist end of the XR spectrum are VR worlds whose structure and content aren't built to have any relationship to the user's real world surroundings. Most VR applications today work like this though we can imagine VR simulations that might fit closer to the middle ground of the XR spectrum. For example, in Chapter 4 we looked at an early VR study conducted by David Paul Pertaub and his colleagues (Pertaub, Slater, & Barker 2002). Pertaub built a seminar room in VR that was meant to look a lot like the real-life seminar room subjects were in when they took part in the study (in that sense while the VR world does not overlay content onto an existing world, the VR space is built to simulate the world subjects were in as closely as the technology of the time allowed). It's possible, indeed highly likely, that VR and AR technologies will merge in the future to maximize user choice and streamline experience (i.e., where one and the same piece of hardware can give users experiences all along the XR spectrum depending on the application and user preference).

Let's take a look at a few real world AR applications to help us see how these technologies might continue developing in the not-too-distant future. These examples can help us see that the issues I raise in this chapter aren't merely hypothetical. The seeds of these problems are already planted. As early as 2007 a Canadian company called Modiface began marketing itself as a developer of "augmented reality technologies for beauty brands" that include "photo-realistic makeup simulation with dynamic lighting adaption, photo-realistic hair color and style simulation" ("Modiface"). AR like this lies on the minimalist end of the XR spectrum but it's instructive for us to see how such AR applications give us a vision of the future of AR too. Modiface's AR applications make sense as natural extensions of current "try before you buy" makeup counters. From a capitalist corporate point of view, the attraction of applications like the kind sold by Modiface make even more sense: you can give consumers a chance to try your product without actually using any of it! It also makes sense from a consumer perspective: if it turns out that you don't like how you look in AR at least you don't have to remove the makeup from your actual face. We can extrapolate a little further and imagine a future where AR simulation is omnipresent and so AR makeup just *is* what users choose to wear.

Social media applications like Tik Tok (ByteDance 2019) and Snapchat (Snap Inc. 2011) already include AR overlays that users employ to enhance and modify their videos and photos, and third party applications allow for even more variety. These overlays themselves range from the humorous (adding canine features to a user's face) to overlays intended to be invisible to others (e.g., body sculpting filters intended to make users appear thinner). AR overlays are also used by surgeons, including cosmetic surgeons, to help patients not only understand the nature of their procedures but also to see "before" and "after" images overlaid onto their bodies (Fallahi et al. 2021). Before looking at the real-world effects these AR applications are already having on users, I want to first conduct a brief review of (physical) body modification in general and the ethical literature on body modification that has developed around the practice.

Autonomy, Representation, and Body Modification

AR overlays already let us change our appearance in ways both subtle and radical but this shouldn't, and probably doesn't, surprise us. People have used the technologies available to them to change the way they look arguably for the entire life of our species (Danesi 2018). In the modern Western context, bioethicists have wrestled with competing theories about when, and why, body modifications are morally acceptable. This is important when we consider the wide range of body modifications people choose to undergo today without raising ethical red flags. For example, what clothing we choose to wear, whether to use (or not to use) makeup (and how much), to get or not get a tattoo (and where), surgical augmentations of all sorts, surgical removals of various kinds,[5] and still other surgical procedures intended to alter the human form in more unusual ways (e.g., forking a tongue, inserting steel implants under the skin to create horns, etc.).

Traditionally, bioethicists analyze the ethics of invasive procedures like these in terms of three sets of considerations: respect for individual autonomy, a duty not to harm (beneficence/non-maleficence), and considerations of justice (how are scarce medical resources distributed and why?) (Gillon 2015). When it comes to questions about when body modification becomes unethical, considerations of autonomy are especially important (Culver & Gert 1990; Culver and Gert 2004; Nelson & Ramirez 2017). Perhaps the most common example of a situation where autonomy and body modification come into conflict is in the realm of cosmetic surgery. This is especially true for patients who undergo repeated cosmetic surgery; this ethics of body modification of this kind is often complicated because such would-be patients are sometimes suspected of having body dysmorphic disorder (BDD) and, if so, their requests for modification are usually refused on that basis (Higgins & Wysong 2018).

Body dysmorphic disorder "is a psychiatric disorder characterized by preoccupation with an imagined defect in physical appearance or a distorted perception of one's body image" (Higgins & Wysong 2018). Because individuals thought to have BDD are in the grip of a specific kind of delusion about their bodies, their autonomy is often considered compromised. A surgeon's duty to respect patient autonomy and their duty to avoid harming patients thus leads them to side against going through with the surgery. Generally speaking, it's typically assumed that individuals in the grip of mental illness are not rationally autonomous when they express desires grounded in their mental illness (Nelson & Ramirez 2017). Note that the reason why ethical issues exist when it comes to patient requests for body modification in instances where mental illness is suspected has *little* to do with the *kind of request* a person makes but instead with the presumed (medical) competence or incompetence of the patient coupled with the fact that surgical operations are not easily reversible. If a patient is autonomous and the surgical modification being requested would not be life threatening, requests for body modification are usually permissible even when they run counter to social norms about how people "should" or "should not" look. This fact will be important as we consider the ethics of AR embodiment and augmentation.

Another important aspect for us to keep in mind is the concept of representational intersectionality (Crenshaw 1991). I introduced this concept in Chapter 5 as part of an argument against empathy-enhancing VR and AR simulations. There it was introduced to help us get clearer on how structural forms of intersectionality (i.e., the view that internalized norms and concepts associated with our identities can impact how we experience the world) differed from other intersectional frameworks (e.g., political and representational). Here I reintroduce representational intersectionality as a reminder that how we choose to represent people (especially how we represent bodies as raced, classed, gendered, and abled) is a complicated process with an ethical dimension.[6] Body modification in general is usually seen as morally permissible for someone to do so long as the person looking to modify themselves is judged as being autonomous and medically competent enough to understand and appreciate the risks involved. Although the ethical landscape for physical body modifications, while contentious in many cases, is fairly well understood, AR body modification introduces new dimensions into this analysis. Let's jump into those issues next.

AR Embodiment and Filter Dysphoria

We don't live in a world of omnipresent AR meta-informational displays, at least not yet. Today we interact with AR overlays while playing mobile-based AR games like Niantic's *Pokemon Go!*, sending friends a photo using AR filters, consulting with an interior designer about home remodels (Siltanen, Oksman, & Ainasoja 2013), shopping for makeup, or consulting with cosmetic surgeons. When it comes to issues of the body, we're already seeing some effects of AR on how users feel about their own unaugmented bodies.

Social media applications like Snapchat and videoconferencing software like Zoom have both been associated with causing user unhappiness with their unaugmented bodies; these effects have led some psychologists to express concern over Snapchat and Zoom dysphoria (Ramphul & Mejias

2018; Rice, Graber, & Kourosh 2020). Because our AR overlays are not fixed, because AR is not yet omnipresent, our AR bodies and our physical selves can look, sound, move, and feel very different from one another. In this case, it seems as if AR technologies are accelerating an already dangerous increase in body dysmorphic disorders that result from the combination of harmful gendered beauty norms and a technology that allows users to modify their bodies in AR to better meet those harmful norms.[7] In a world of omnipresent AR, a world where AR overlays can follow users throughout their interactions, this form of dysphoria may cease to exist since the mismatch between the AR augmented self and the everyday self would, or at least could (if the user chooses), disappear.[8] That, at least to some, may seem like an advantage for omnipresent AR. It also seems like a natural extension of our current norms over autonomy and body modification. If, at least for autonomous rational agents, it's morally permissible to use clothing, makeup, pharmaceuticals, tattooing, and surgical procedures to modify bodies then it's hard to see a principled reason to deny those same agents from augmenting or modifying their AR bodies.

This seems even more true in a world where AR simulations have been integrated into the basic structures of our society in the same way that the internet has become integrated. Although it's possible to live without internet access today, doing so comes at steep personal and professional costs. For example, for philosophers (and for most academics), job postings are almost entirely and exclusively available online. Similarly, almost all academic research is submitted and vetted using online manuscript portals (long gone are the days when academics would send their papers and books to publishers using the postal service). An academic not able to access these job postings would be at a monumental disadvantage relative to those who do have access. I'm envisioning a 2040 where AR has come to occupy a similar place in our personal and professional lives (where things like job postings, personal and remote conversations, and so on all happen with AR overlays). As we saw in the previous chapter, professional codes of ethics like the one offered by the Association for Computing Machinery (ACM 2018) argue that technologies that stand to be integrated into the basic structure of society deserve the strictest forms of ethical scrutiny.

Despite the fact that using AR to modify the self is both a natural extension of our current practices governing the ethics of (physical) body modification and because of the near certain development of future AR systems that allow (and encourage) the practice of AR body modification, we must carefully examine the ethical and regulatory issues that these technologies will introduce. In the following two sections, I'll argue that AR body modification introduces two connected issues we aren't currently equipped to handle and which, at the moment, don't admit of easy solutions. The first issue centers on who should have the right to control user representations while the second focuses our attention on the fact that the epistemologies of reidentifaction (i.e., the tools we use to keep track of ourselves and other people) are not ready to handle a world where identity and the body are completely severed from one another.

Issue #1: Who Should Control How People Appear to Others?

The first issue we need to address in a world of omnipresent AR requires that we answer a question about who should have control over how people appear within an AR space. In this section I survey four natural-seeming responses to that question: users should control how they appear to others, users should control how others appear to them, or that states (or corporations) should provide regulatory (or policy) frameworks that set boundaries on AR appearance. In the context of our discussion, I'll borrow a distinction introduced by philosopher Erica Neely regarding what she calls public and private AR spaces (Neely 2019).

Neely focuses her analysis on AR *property* and thus on the modification of spaces (homes, offices, parks, etc.) and not on AR bodies and AR body modification. Her analysis of the ethics of AR overlays turns in part on a distinction she draws between private and public AR spaces. Importantly, it also turns on the actual means by which users will come to experience AR in the future. For example, she considers the fact that

> there could be a multitude of different apps which augment content; we would use different apps to experience different content. Second, augmented reality could be much more seamlessly integrated into our experiences, where users either do not have to consciously choose to experience the content or where users have converged on the same platform for content such that there is essentially a single augmented sphere that people access.
>
> (Neely 2019)

In this section, I'm going to explore the ethics of AR embodiment from the point of view of Neely's second option. It seems likely *both* of Neely's options will see development in the future. Current AR applications (*PokemonGo!*, Modiface's AR makeup counters, etc.) are like Neely's first form. Their AR overlays are not accessible to everyone but only through users who have accepted some kind of licensing agreement with an individual company. While I think the ethics of these sorts of arrangements, and the sort of content that may or may not be ethical in such spaces, is worthwhile, many of those questions are already covered by the code of ethics I offered in Chapter 6. [9] Much less has been said about the ethics of omnipresent AR spaces and AR embodiment. All of the options I'll survey here have a degree of plausibility and I don't intend to give an argument in favor of one response over another. What I think we can begin doing here is to assess the relevant moral variables involved in choosing one response over another (or some combination of responses) because each will introduce risks or tradeoffs.[10] It seems natural to assume that AR will develop in the direction of creating a shared common space for users to inhabit. This is true not only because unaugmented physical spaces are also shared (and AR overlays information onto those already shared spaces) but because of the importance of our external physical presentation to our sense of who we are. It's also natural to assume that people will want to exercise some control even in AR public spaces over their appearance (much as they do in today's unaugmented spaces).

The first way of answering the question about who should be able to determine how we look in AR is to say that individual users should be able to control how they appear to other people. It's natural to assume that we, that is, that *you* should be the one to control how your AR body looks not only to yourself but to people who share an AR space with you. This is true not only because it follows naturally from the fact that you already exercise the most control over how your physical body looks but also because that answer seems to also be the conclusion of our conversation about the ethics of physical body modification. Autonomous (rational, medically competent) people can and should be allowed to modify their bodies as they see fit so long as they understand the risks associated with the modification. This holds true for small (wearing green today instead of blue) and large-scale modifications (whether to undergo reconstructive breast surgery immediately following mastectomy) (Hansson, Elander, Hallberg, & Sandman 2020).

However natural this response may seem, it isn't without cost. As anyone who's been online has learned, "anonymity leads to more trolling" (Guo & Caine 2021). If users are empowered, without any limitation, to control how others see them in a shared AR space then we may inadvertently create an AR domain that mimics the very worst aspects of anonymous online trolling environments. Trolling, in this context, can be defined as any behavior that "involves deliberate, deceptive and mischievous attempts to provoke reactions from other online users" (Golf-Papez & Veer 2017). If we opt to empower users in this way, there appears to be nothing stopping users from donning digital blackface or parading with others in a march of Hitler clones just to create chaos. Worse still are trolls intent on causing trauma to specific people.

On October 1, 2013 Jerry Brown, then governor of the state of California, approved Senate Bill No. 255, paving the way for California to become the first state in the US to specifically outlaw "revenge porn" (SB 255 2013).[11], [12] Although laws like this may prevent users (or, at least, legally punish users) from modifying their AR bodies to enact "revenge porn" against an ex, the laws on simulated but realistic nude images, known as deepfakes, are currently unclear throughout most of the world. Despite revenge porn laws, which would prevent a user from choosing to use actual nude photos of someone else to overlay onto themselves in AR, realistic, but nonetheless entirely simulated, nude images are harder to regulate (de Ruiter 2021; Johnson & Diakopoulos 2021; Westerlund 2019). Additionally, such laws would not prevent users from using images of notorious historical persons, like Adolph Hitler, as their AR persona nor would it prevent individuals from donning digital blackface.[13] Setting rules, regulations, or company policies that would prevent these forms of trolling would require siding against user autonomy when it comes to the power to control how others view one's AR body. While I'll consider these state and corporate options, I want to first consider the consequences of another option: allowing users to control how others appear to them.

This option also appears to flow naturally from pre-existing consumerist principles. Users, understood as consumers of content, are usually allowed wide latitude in terms of their freedoms to choose what information they expose themselves to. Aside from strictly illegal activity or content, for example, users in most countries are free to visit almost any website they choose but they usually are not legally *required* to visit specific sites.[14] Reasoning along these lines may lead us to conclude that users should be allowed the freedom to choose how others appear to them in the context of omnipresent shared AR spaces. Perhaps today I'm in the mood to see everyone around me as my favorite cartoon characters or maybe I'd rather not see AR embodiment overlays at all. Shouldn't users have this level of control over their access to AR content? This too seems like a plausible response to the question of who should have control over AR representations of people. However, as with the prior response, empowering users in this way comes with costs of its own.

First, and most obvious, empowering users in this way *disempowers* other users by stripping them of control over their physical representation. In our unaugmented world, while you're not required to look at me when we're in a shared space, you're also not allowed to force me to look a certain way (assuming that I'm not breaking any local laws). However, some readers may think that perhaps this tradeoff is worth considering despite this downside. I think that there are other problems with granting individual users the power to see others as they wish. Deepfakes, for example, emerge again as a problem. Deepfake technology "refers to machine learning techniques that can be used to produce realistic looking and sounding video or audio files of individuals doing or saying things they did not necessarily do or say" (de Ruiter 2021) and it is becoming increasingly difficult to tell the difference between genuine photos and videos of a person and a deepfake photo or video.[15] Projecting forward into our imagined 2040, we can assume that such images and videos are indistinguishable from real images and videos and that they can be processed in real time. As a result, it's entirely possible that users, empowered to choose how others appear to them, would choose to use deepfake technology to see everyone around them naked.

This sort of technology might also be used in other equally problematic ways. For example, a committed racist may understand that their racism is unpopular and so choose to use AR to have all people appear to them as members of their preferred race in order to keep them from being overtly racist.[16] Ethicists who think about deepfakes sometimes argue that "our image and voice are closely linked to our identity" and that because of this "protection against the manipulation of hyper-realistic digital representations of our image and voice should be considered a fundamental moral right in the age of deepfakes" (de Ruiter 2021). Although this argument may make sense for us (those of us living in an unaugmented world), it makes increasingly less sense in a world where the ties between our physical bodies and our AR appearance are more loosely connected (if at all). I'll argue in the following section that, in many parts of the world, cultures are re-examining the presumed link between our bodies and our selves. If I'm right about this then it's not clear whether the arguments that speak against deepfaked nudity today will make much sense in the future. For now, I think it's enough to say that empowering users to decide how others view them or empowering them to decide how others look to them both introduce problems we're not currently equipped to handle.

Readers may notice that both of these options, in an important way, only make sense against a backdrop of enforcement. Even if we think one of these first two options is *the right option*, it isn't possible to implement these options without the backing of either state or corporate power. I'll briefly look at these two options next. What happens, for example, if we empower states to create regulations about how users can represent themselves (or view others) in AR spaces?

This option is also a natural extension of how things work in most countries today. We are free, within legal limits, to choose what to wear (and in some places, free not to wear anything at all) before heading outside our private spaces. Why shouldn't an AR space merely adopt this attitude as well? On the one hand, this approach has the advantage of not really requiring much change. Whatever laws currently exist governing decency and dress in public will apply in AR spaces as well. On the other hand, we have many more options available to us in AR about how we look than we do in unaugmented spaces and our current regulations probably won't adapt to these cases very well. For example, while someone may be able to make a living as a celebrity impersonator today, they'll always be limited by their bodies in terms of how well they look and sound like the celebrity they're impersonating. AR eliminates this barrier. It's not clear that laws are currently equipped to deal with people who choose to look like non-humans in AR (e.g., someone who chooses to look like a tree to spy on people without concern or who chooses to look like a deer while out in the woods). Similarly, as I mentioned above, our current regulatory frameworks are ill equipped to deal with deepfaked representations in general. Some readers may also worry that empowering states to control appearance is a step too far toward representational authoritarianism.

State use of AR can easily capture the dystopian imagination. For example, a 2014 installment of the British program *Black Mirror* envisioned a future where governments use their control over omnipresent AR embodiment to digitally shun citizens that step out of line (Brooker & Tibbetts 2014). In an episode, titled "White Christmas," Matt (played by Jon Ham) is digitally "blocked" after being convicted as a sex offender. Being blocked, in this world, means that others appear to Matt only as silhouettes and the blocked user appears as a silhouette to everyone else. Blocked users are not able to hear others and are, at the same time, unintelligible to unblocked users. If access to AR spaces is both omnipresent and vital, much as internet access is for many today, then such punishments are not only within the capability of a state-level organization but they're powerful punishments indeed. We should be cautious about empowering states to control AR embodiment to this degree (though, all things considered, it may be that this option is ultimately worth pursuing). Concerns like these may lead people to explore yet another way of addressing issues of control and representation: corporations should regulate people who use their AR applications.

As with the three other options explored in this section, there's something natural about the claim that corporations who develop AR applications should be responsible for laying out rules for their users and that it's these companies that should set the bounds of proper, and improper, forms of AR embodiment. When Twitter, and then later Facebook, chose to ban then US president Donald Trump from their platforms, they were exercising a form of control over how Trump was allowed to represent himself on their platforms (Fung 2021; Cillizza 2021). Similarly, author Naomi Wolf was banned from Twitter in 2021 after posting tweets containing misinformation about the Covid vaccine (Wolf 2021). In these instances, each corporation decided that users had violated their terms of use and, as a result, took steps to remove the user from their platforms. In Trump's case this process took years whereas Naomi Wolf's ban happened very quickly.

These examples show us both reasons for and against giving corporations the power to decide the rules under which users can embody themselves on AR platforms. Philosopher Luciano Floridi, analyzing the decision made by social media corporations to ban Donald Trump, came to the conclusion that this decision, while it was the right one, also gives us reasons to think that online spaces should be treated as Commons and not as privately owned spheres. Online social space, he argued,

> is the space where humanity spends more and more time and where more and more activities take place directly or indirectly, from education to work, from socialisation to entertainment, from commerce to finance, from the exercise of justice to political discussion, from research to journalism. It is the space that influences every other space, even the physical one; just think of all the issues surrounding defence and security. It is a space that should be conceptualised and governed more like a condominium —like Antarctica and the Space Station, which belong to everyone—rather than like a new frontier that can be appropriated and colonised by anybody, or like a space that belongs to no one, like the Moon.
>
> (Floridi 2021)

If corporations were fully empowered to control who can, and who cannot, embody themselves in AR, or if corporations are empowered to enforce their own rules about representation, then we are likely to see similar problems in the future. Users who stand to draw attention, views, the AR-equivalent of clicks, notoriety, or who might increase the user base for a platform are likely to receive a lighter touch from corporations when it comes to enforcing their terms of service. It's been argued, for example, that despite repeatedly violating Twitter's terms of service, Donald Trump's account was allowed to remain in operation because it received special treatment from content moderation teams who decide how to apply their harassment rules (Mac 2019). Social media CEOs have, in some cases, even admitted that their platforms are not set up to deal with abusive users (Tiku & Newton 2015).

Recall that the code of ethics put forth by the Association for Computing Machinery lays out special obligations for developers of systems that become integrated into a society's basic infrastructure. This code cautions developers that

> [w]hen organizations and groups develop systems that become an important part of the infrastructure of society, their leaders have an added responsibility to be good stewards of these systems. Part of that stewardship requires establishing policies for fair system access, including for those who may have been excluded.
>
> (ACM 2018)

In the case of AR embodiment, it's important that we take seriously Floridi's argument that spaces like these can not only become places where we spend more and more of our time but also spaces where we carry out our most basic commercial, social, and political activities. As such, it's important that access to such spaces, including the rules that govern their use, is distributed fairly and that rules are enforced consistently. A strictly corporate response to the question "who should control how users are embodied in AR?" would have a tough time accomplishing this task given the profit incentive behind allowing popular, but bad, actors to stay on a platform. In addition, some readers may be wondering why corporations should control something as intimate and personal as our embodiment given how closely connected this is to our sense of who we are and how we express ourselves.

Throughout this section, we've been looking at four natural-seeming answers to questions about who should have control over how we represent ourselves in AR spaces. This question is likely to become increasingly important as we spend more of our time in simulations across the XR spectrum. The temptation to overlay simulated content not only onto the world (as Neely explores) but onto our bodies is both natural and inevitable. The question about who should control how users in AR spaces appear to one another is equally inevitable and the four most natural responses each introduce problems we're currently not ready to face. Although I've presented these options as mutually exclusive, any answer we choose will require some degree of state-backed regulation.[17]

This fact not only further complicates the ethics of AR embodiment, it also introduces another related set of issues regarding the epistemology of identity and reidentifcation. No matter how we answer the question about user representation, we'll also need to be able to keep track of users over time. We'll need this ability if only to make sure that justice is carried out, that citizens don't vote more than once in elections, that money is drawn from the right accounts when purchases are made, and that we're spending time with our actual loved ones and not AR pretenders. Some system, in other words, needs to be in place to help users keep track of who they're spending time with. Just like our first issue, there are some natural-seeming ways of solving this problem and, just like that issue, these natural-seeming answers introduce new problems we'll need to be ready for.

Issue #2: How Do We Know Who We (or Others) Are when Our Evolutionary Heuristics Fail Us?

To see why we need a system for keeping track of identity in AR spaces, it might help to think about how we reidentify people today. Earlier in the chapter I noted that whatever heuristics we use (in physical space and in AR spaces) need to be sticky and stable. Stability, in this context, means that we need to be able to reliably track the features across situations. Stickiness refers to the ability of our heuristics to track features that reliably pick out a distinct person. For example, someone being tall might be stable (I can track height across most situations) but is not very sticky (there are a lot of tall people – at least where I live). When I met my friend for coffee, I relied on a whole suite of multimodal tools designed to work together to give me a sticky and stable way to make sure that I got up to greet the right person (and not a stranger). Social psychologists have long known that "[t]he accurate perception of others is a fundamental aspect of social cognition, allowing us to detect the intention, attention, and identity of an individual (among other attributes)" (Maguinness & Newell 2014). How do we do this in an augmented world?

Given the importance of social cognition, it shouldn't surprise us to learn that we've developed heuristics designed to help us recognize ourselves and other people. Perhaps most obvious for sighted individuals are heuristics oriented around faces. Psychologists have repeatedly discovered that "face recognition abilities are rapidly acquired and can be influenced by our surrounding environment" (Maguinness & Newell 2014).[18] If I ask you to think about yourself, an old friend, your children, or your spouse(s) it's likely that you'll do this by forming a mental picture of that person/people and that this mental image prominently features their face(s). We seem *built* to use faces for reidentification. Neuroscientists have even suggested that we've evolved a brain region that appears specifically built to identify and recognize faces called the Fusiform Face Gyrus (FFG)[19] (Kanwisher, McDermott, & Chun 1997).

> [R]esults confirmed the partial distinction of the brain mechanism involved in recognition of personally familiar faces and that in recognition of one's own face. The right occipito-temporo-parietal junction and frontal operculum appear to compose a network processing motion–action contingency, a role of which in visual self-recognition has been suggested in previous behavioral studies. Activation of the left fusiform gyrus selective to one's own face was consistent with the results of two previous functional imaging studies and a neuropsychological report, possibly suggesting its relationship with lexical processing.
>
> (Sugiura et al. 2005)

However, our ability to recognize ourselves and others isn't entirely dependent on needing to recognize faces. Research suggests that we can recognize (and reidentify) our friends just by looking at silhouettes of them walking (Cutting & Kozlowski 1977), by the sound of their voices (Kaplan et al. 2008) and that we even seem to have evolved an ability to recognize our own distinct smell (Milinski et al. 2013).[20] It's important to us for evolutionary (and metaphysical) reasons that we be able to keep track not only of ourselves and our loved ones but also that we're able to recognize people from one encounter to another. If we weren't able to reward those who cooperate with us and sanction those who take advantage of us, it's not likely that we would have survived long enough as a species to develop VR and AR technologies in the first place!

The irony is that these very technologies now threaten the mechanisms we've developed to let us engage in social recognition. In Chapter 6 I referred to this idea as "deskilling" (Vallor 2015). Deskilling refers to the process of weakening (or losing) a capacity as a result of technological developments. Our abilities to engage in neurotypical social cognition/social epistemology are likely to be deskilled in a world of omnipresent AR embodiment. These abilities, whether it be our special attunement to faces, our ability to distinguish our smell from others' (or the sound of our voice or our unique movements) all depend on a tight-knit connection between our physical bodies and our selves. Because AR overlays can weaken or even eliminate this connection, we'll need to develop new ways of keeping track of our selves and of each other in order to retain the benefits of social cognition in an AR landscape.

At this stage it's worth clarifying the issue I'm examining. When philosophers ask questions about personal identity and the reidentification problem they're usually speaking about a slightly different set of questions than the ones I'm looking at here. Traditionally, personal identity is understood as a question about *metaphysics* (Parfit 1971; Schechtman 2005). In other words, what *is* a self? Do enduring stable selves actually exist over time or is this an illusion? If they do exist over time, what lets them do that? These questions demand a metaphysical answer. They cannot be answered without providing a theory about what people *really* are and not just what they seem to us to be.

In this chapter I'm not trying to answer *those* questions.[21] Instead, I'm asking a set of questions that philosophers would refer to as *epistemological* questions about identity. Epistemology is a branch of philosophy that deals with questions about knowledge. Instead of asking what selves really are (and what they're not), in this chapter we're asking questions about how we come to *know* (to form a justified belief) that we're meeting up with our friend at the coffee shop and not a stranger (or how we keep track of our own identity over time). These questions are related of course (e.g., if it turned out that enduring selves are an illusion then that conclusion has major implications for an epistemology of selves). I've surveyed some of the evolutionary heuristics that we've developed to address these epistemological questions (i.e., we keep track of selves by looking at facial structures, gaits, odors, and sounds that help distinguish a person from others).[22] The AR landscape we're imagining, however, is one where our heuristics break down. In that sort of world we need new epistemological heuristics to keep track of identity over time. What heuristics might we develop to help us do this?

Our options for addressing this question are less obvious than for the first issue I introduced about identity and control. I'll look at three ways we might keep track of identity in a world where the way people look, sound, feel, and even the way they move is severed from how they're physically embodied. I'll begin with a critique of the problem itself. Why not just do what we already do and use some kind of biomarker to keep track of people or, better yet, why not just say that your "real self" is whomever you are when we turn the AR overlays off? I'll then survey two options less committed to contestable claims about the body and its relation to identity. The first will be to consider creating Non Fungible Tokens (NFTs) to keep track of user identity over time. The second approach answers questions about identity by generating inferences about identity from patterns of user activity. We are, on that option, a reliably occurring pattern of behaviors and interactions. I'm not sure how best to respond to this question but it is a question we'll *need* a response to. Let's begin by looking at the problems with using biomarkers as a way of keeping track of identity.

Biomarkers

Perhaps the most obvious response to questions about how we should keep track of our selves over time in a world of omnipresent AR embodiment is to argue that AR embodiment doesn't pose any new problems for us at all. AR embodiment, someone might say, is irrelevant when it comes to questions not only about *who* you are (metaphysical questions) but also to questions about *how* we keep track of people. People are in AR spaces the exact same person they are outside of AR and outside of AR people are, at least for practical purposes, their bodies.[23] This makes it easy to solve the problem because we can either ground AR identity in terms of a person's physical body or instead on some biomarker related to their physical body (e.g., fingerprints, DNA, etc.). This way of thinking is both natural (it makes sense to want to import our existing heuristics onto AR embodiment) and also, probably, mistaken. There are at least three ways in which this way of thinking is problematic and while the fact that it's problematic doesn't mean that thinking of identity in terms of the body/ biomarkers is ultimately the wrong way to go, I think its costs are probably too high.

First, identifying our selves with our physical bodies seems to miss the mark about the nature of the problem. To figure out whether I'm meeting with my friend at the coffee shop, on this view, we would need to ask each other to turn off our AR overlays so that we can see who we're "really" with. This response is akin to telling someone to avoid the internet if they don't like being bullied online. In a world of omnipresent AR embodiment (a world where most of us are nearly always in a space that includes AR overlays) and where existing in AR spaces is both already a basic fact of our lives (i.e., that AR is integrated into the basic structure of society) there are serious costs to tying identity to our unaugmented physical bodies. As with online bullying, removing the relevant technology might solve one problem (e.g., online bullying can't occur directly if one isn't online and questions about AR embodiment can't occur if we're not in AR spaces), but it does so by enacting a deeper injustice. Removing a person's ability to express their identity by modifying their AR bodies is probably an answer of last resort if we're mostly existing in AR.[24] This is especially true in a world in which participation in AR spaces is as important as internet access is today.

As I noted near the beginning of the chapter, modifications to our physical bodies can be plotted along a spectrum that encompasses both very minor (clothing, makeup, hairstyle) and very radical transformation (nullification, animality, etc.). Debates about body modification make clear that the present moment is at least partially defined by a multi-pronged set of challenges to traditional conceptions of the body and its relation to identity. As late as 2012, for example, someone who identified as a sex or gender different from the one assigned to them at birth could be diagnosed with a mental illnesss then known as Gender Identity Disorder (GID) (American Psychiatric Association 2000). Gender identity disorder no longer exists as a diagnosis in the most recent version of the *Diagnostic and Statistical Manual of Mental Disorders* (American Psychiatric Association 2013). This happened, at least in part, because of how psychologists, and the culture more broadly, have come to rethink the nature of transgender identity.[25] A series of political, social, and scientific paradigms have begun to shift towards a more inclusive conception of the self, one where the self (who and what we are) is less directly connected to natal embodiment.

We're in the middle, in other words, of a large-scale shift in public attitudes about not only the medical status of trans persons but also the ethics of trans identities in a much broader sense. Scholars and social activists alike are currently discussing and debating the nature of a wide range of trans identities, including transabled and transracial identities (Barnes 2019; Bettcher 2013; Bogardus 2020; Butler 1999; Fernandes Botts 2018; Haslanger 2000; Hargie, Mitchell, & Somerville 2015; Hilton & Lundberg 2021; Müller 2009; Overall 2004; Sealey 2018; Tuvel 2017). I don't here intend to speak to these specific debates in any detail. I mention them, instead, to provide some evidence for thinking that our concepts about the self and its relation to the body are currently under radical dispute in ways that haven't been seen in generations. One implication for the "identity as unaugmented body" approach to fixing AR identity over time is that it swims against the mainstream current in this respect because it links identity strongly with unchanging aspects of a person's biology.[26] Representing ourselves in AR spaces as being tightly bound to our natal (or even to unaugmented) biology in this way may also create problems from the point of view of representational intersectionality (Crenshaw 1991). How we represent ourselves and others is always an ethical matter. Similar problems would arise if we were to try to sidestep the body and use a biomarker to keep track of identity in AR spaces instead.

What could be wrong with identifying someone with a fingerprint, DNA sequence, or with facial geometry? While biometric identification is both sticky and stable (and so really could be used to keep track of identity in the AR spaces we're imagining), it faces a slightly different set of problems. First, biomarkers like these are, at least today, cumbersome to actually get access to. We don't typically ask for someone's fingerprint or for a DNA sample before greeting them at a coffee shop to make sure that they're really an old friend. Whatever we use to keep track of identity in AR needs to be as seamless as the heuristics we use today. One the one hand, this may only be a minor issue as we can imagine a world where such information is readily, maybe even constantly, being monitored and relayed to other users. Note, however, that the infrastructure required to make this sticky and stable information accessible to people in AR requires that users give up a lot of private information. Not only would users be giving up biometric information to others but they would also allow the relevant corporation (or state or both) to have access to their physical location at all times (this would be necessary so that other users could keep track of their identity).[27] However, biomarker approaches would still face the problem of running counter to the evolving ethics of trans identities. So long as identities are fixed by biology (instead of a trait or feature that users have control over), biomarker approaches to tracking identity will remain problematic.

To avoid these consequences, it would be better to find a sticky and stable way of tracking identities in AR spaces that doesn't rely strongly on increasingly controversial assumptions about the relationship between the body and the self. Such a view, were we able to find one, would be compatible with almost any outcome about debates regarding the fit between the body and the self. I'll explore two approaches that attempt to do this.

NFTs and Identity Hashing

In 2021, artist Mike Winkelmann sold a digital artwork, *Everydays*, at auction for $69.3 million. *Everydays* was sold as a Non Fungible Token (an NFT) to an undisclosed buyer but the sale of *Everdays* was unique not just because of how much someone paid for it (the sale broke records for digital art) but also because of what it was that was actually sold (Moscufo 2021). NFTs like *Everydays* are not unique to art or the art world. In the same year, a short clip of basketball star LeBron James dunking during a game sold as an NFT for almost $400,000 (Young 2021). What buyers purchased in each case was a unique connection to the work though the *way* in which it is unique helps us to see not only what NFTs are but also why they might help us address questions about AR identity.

When an anonymous buyer purchased *Everydays*, they did not buy the sole rights to view or display the work nor did they buy a file that can be used to generate the image (both are widely available online to anyone).[28] Similarly, because NFTs are digital artworks, the concept of an "original" or "first copy" of a work is strained. Instead, in both the art and slam dunk NFT cases, "[t]he winning bidder owns the work in the form of a unique string of code, called a nonfungible token. The piece has no physical presence and will be 'delivered directly from [the artist] to the buyer, accompanied by a unique NFT encrypted with the artist's unforgeable signature and uniquely identified on the blockchain'" (Moscufo 2021).

While some may express skepticism about whether the buyers now own anything of any significance or real value, it's impossible to dispute that these buyers bought *something*. What *Everydays*' current owner owns is access to a unique string of information that was transferred from the artist and which cannot be copied (even though the image itself can be copied infinitely as all digital files can be). This string can be tracked across owners and cannot be forged. All of this is achieved without government or state intervention. NFTs seem to have the kind of stability, in other words, that makes them good candidates for keeping track of identity in a world of omnipresent AR. Before I say more about this possibility, just what *is* an NFT?

> An NFT is a blockchain-recorded right to a digital asset. This can be anything digital; an image, a video, a song, a digital trading card of your favourite baseball player, a coded piece of virtual land, or a virtual tunic for your virtual character to wear while he explores his virtual land.
>
> (Dowling 2021)

Blockchain technology (BCT) is a fundamental aspect of what makes NFTs possible. Without BCT NFTs could not exist. Given the importance of BCT, let's be clearer about just what this kind of technology does:

> [B]lockchain-enabled distributed ledger technology is a system of decentralized accurate transaction ledgers that uses preprogrammed algorithms, protocols, and state-of-the-art cryptography to reach an automated consensus about the existence and evolution of transactions. It provides an immutable history of transactions that is transparent to all actors, without the need for a third party to ensure trust between exchange partners by monitoring and validating transactions.
>
> (Schmeiss, Hoelzle, & Tech 2019)

One advantage of BCT is that it's very useful for tracking the history of transactions in a way that is, ostensibly, irreversible and, proponents argue, very secure. When we describe an NFT as "strings of code" what's meant is that the string has a stable identity within a blockchain of transactions. This blockchain is constantly being verified by multiple computers around the world which helps to guarantee the integrity of the blockchain and protect it from tampering. The buyer of *Everydays* now owns the particular (unique) string that represents the transaction itself (an interaction between the artist and the buyer). Thinking of identity in terms of NFTs would require important changes to our current heuristics for identity.

First, each user would be responsible for choosing their NFT and linking it to a key that they can use to access it. An NFT, as we saw with our examples, could be literally anything. What's important is that each user is empowered to choose something (an image, an audio file, a favorite color, a recipe, etc.) which could then be hashed (algorithmically transformed into a unique "string of code") that could then be tracked via BCT. In order for the identity-as-NFT approach to work, all users of AR technology would need to have constant access to a globalized identity blockchain that tracks both a user's location and the NFT associated with that user's key. All users of AR spaces who want to overlay information over their physical bodies would also, we might presume, need to have selected an NFT that can be tracked in the same way. Because NFTs are stable (their major selling point is that they can be reliably tracked over time across multiple transactions), they satisfy one of the requirements needed to track identity across AR spaces, though, importantly, BCT is not without its limitations and security risks (Li et al. 2020).

Although NFTs are stable it's likely that NFTs cannot be made sticky. The current owner of *Everydays* is free to transfer the work to anyone they wish, instantly, by transferring the key they received from the artist. Because BCT is anonymous by design (NFTs, Bitcoin, and so on can be tracked across transactions but the users involved are not) and because transactions of NFTs, unlike transactions in a traditional securities market, are neither regulated nor tracked by states, the identity-as-NFT approach is open to unacceptable consequences. For example, a single person could have multiple NFT identities either through legitimate trading or the use of threats/ force. If a user can easily switch their identities by switching the keys they use to access an NFT then this lack of stickiness defeats the purpose of turning to NFTs in the first place. The opposite problem can also arise. It may be possible for a user to lose their NFTs (either via trading, coercion, or simply losing access to their key).[29] Finding a workaround, one that ties NFTs to particular users to make them sticky, would also defeat the purpose of turning to NFTs in the first place (short of falling back onto bio-markers, it's not clear how we would do this).

It is also worth noting that BCT requires enormous amounts of computational resources and that such resources have an environmental cost (Truby 2018). It's been estimated that the carbon footprint of the computer hardware required to maintain the integrity of today's Bitcoin blockchain (an activity known as "mining" in cryptocurrency circles) approaches the carbon footprint of Ireland (Mora et al. 2018). Such effects are likely to grow *exponentially* if BCT is used to track billions of NFTs (possibly tens of billions of NFTs) to track user identity across a lifetime. This cost may be worth paying but only if NFTs can be made sticky and stable without compromising their main advantage (keeping BCT activity anonymous while identifying users without reference to biomarkers).

Activity Profiles as User Identity

There's at least one more proposal on the table for keeping track of user identities in a world of omnipresent AR. This proposal, grounded on work already used by some law enforcement agencies, was spearheaded by Jessica Su and Ansh Shukla at Stanford University and Arvind Narayanan at Princeton University (Su et al. 2017). Su et al.'s research investigated whether it was possible to "de-anonymize" online activity beginning from the premise that

> each person has a distinctive social network, and thus the set of links appearing in one's feed is unique. Assuming users visit links in their feed with higher probability than a random user, browsing histories contain tell-tale marks of identity.
>
> (Su et al. 2017)

Think, for example, about all of the websites you visit on any given day. Think also about the kinds of links you tend to click on. If you're like most people, you probably visit a fairly small number of sites regularly. Additionally, it's probably true that your social media circle, especially those friends and colleagues whose links you interact with frequently, is equally small (no matter how many friends we have online, most people only regularly interact with a small number of them). Su and her colleagues were able to use these facts about online and social media activity to track anonymous users and, with more than 70 percent accuracy, connect their anonymous activity back to specific users (Su et al. 2017).

While the thought of government agencies having the power to track even anonymous activity back to specific users may be a cause for concern, in our case such a technique may be exactly the kind of new heuristic tool we need to solve problems about AR embodiment. For example, on this view, let's call it the "identity-as-activity" view, user identities would not be dependent on physical location or on biomarkers. Instead, something much more characteristic of us (our online activity) is used to keep track of identity. These seem like definite advantages over the biomarker approach. Compared with the NFT option (where an NFT, once created, cannot be changed), there is more room for individual freedom in the creation of an identity. Additionally, because identity is tracked heuristically by looking for reliably correlations of activity, these correlations can't really be traded or sold in the way that NFTs could be. Despite the similarities that might exist between any two users, it's unlikely that two users will have literally identical patterns of online activity. In this sense, the "identity-as-activity" heuristic is sticky (at least sticky *enough* for most practical purposes). For these reasons, this heuristic may have more going for it than either the biomarker or "identity-as-NFT" heuristic though it is not without costs of its own.

One cost to thinking of identity as activity may be fairly obvious. If we think about our online and social media activity as it stood ten years ago it's probably different from the profile of our activities today (possibly *very* different). This heuristic may be good at tracking identity in the short term (while a pattern of user activity remains somewhat stable) but may flounder when it comes to tracking activity over a longer timescale. Similarly, users may either deliberately choose to radically alter their patterns as a way of throwing off the algorithms used to track correlations of their activity or they might manifest very different patterns naturally depending on situational features (i.e., our activity at work may be very different from our activity at home or on vacation or while staying with family, etc.). The heuristic may have trouble tracking us if it turns out that our patterns are less coherent than we think.

I'll pause here to note that while this means that the "identity-as-activity" view would be sticky but not stable, this may be read by some readers as not a cost at all but instead as a more accurate reflection of how identities really do work (metaphysically). For example, the Buddhist doctrine of anattā, sometimes referred to controversially as the "no-self" view, is a rejection of the idea of a stable self that endures throughout a lifetime (Rudd 2015; Sugunasiri 2011). For readers who hold such a view of the self, this heuristic can be seen as better reflecting the fact that our selves really do dynamically change not only over time but also across situations. For readers more inclined to think of the self as something that endures over time despite radical changes in our beliefs, bodies, and social media activity, however, this heuristic and its lack of stability is a serious cost. This is compounded by the fact that, practically speaking, we need a way of tracking user identity in omnipresent AR spaces that allows us to buy things, vote, and so on. Such activities require more stability than can be offered by this approach. It's possible to imagine keeping track of patterns of user activity over time in order to avoid this problem, though doing so would also come at a cost. Some organization or collection of organizations (states, corporations, or global organizations of some sort) would be needed to keep track of this activity and this comes at a cost to privacy. We may not want to have *all* of our activity tracked.

Looking at all three options for new heuristics to track identity it's not clear which (biomarkers, identity-as-NFT, identity-as-activity) gives us most of what we want out of a heuristic while sacrificing the least. Any heuristic that requires locational tracking (which arguably all of these options do) sacrifices a degree of privacy many of us grew up expecting (though any of us who constantly enable location tracking on our smartphones will have already grown to accept this loss of privacy). It is possible to imagine creating spheres of privacy within each of these heuristic options. Erica Neely, for example, usefully distinguished the ethics of modifying private versus public AR spaces (Neely 2019). Applying that idea to our context we can imagine a heuristic that provides varying degrees of information to other users about who we are depending on, whether we're inside a private space (anyone I've allowed into my house gets a high level of access to my information) or a public space (people in public not on a pre-approved "friends" list may get only very basic information about where I am).

I don't propose to have settled these issues to anyone's satisfaction (including my own). What I've tried to do in this chapter is suggest that we have a looming set of problems on our hands associated with AR embodiment. First, that the system of norms we've already developed around body modification will translate naturally toward modification of AR bodies. Second, AR will naturally become integrated into the basic structure of society in the same way that the internet has become. Third, this level of integration will require us to address questions about who should have control over how users appear to one another in AR spaces and, last, that the heuristics we've naturally developed to track identities over time won't work in a world of omnipresent AR. We'll need new heuristics in those spaces and the options all have costs. I end the chapter, and the book, by examining two corollary issues related to this conversation.

Relationships

Now that we have a better sense of the questions we'll need to answer once AR embodiment becomes common, it's important to see how other concepts might need to adapt as well. Scholars have been investigating "online" relationships for decades (Cooper 1999; Cornwell & Lundgren 2001; Munn 2012), but relationships in VR and AR are largely unexplored and under-theorized. Because of the possibility of virtually real experiences in such spaces, it's likely that relationships in VR and AR can, for that reason, look (and feel) a lot more like relationships in unaugmented spaces. However, although virtually real experiences promise to bring XR relationships closer to our real-world relationships, AR embodiment confronts us with a set of issues that cut to the heart of our unaugmented concept of relationships (especially, though not exclusively, our concept of romantic relationships).

The specific nature of the issues arising from AR embodiment will partially depend on how we answered questions earlier in the chapter. For example, one set of issues will turn entirely on whether we believe that users should control how they appear to others in AR or whether users should control how others appear to them. Imagine, for example, a couple spending time together in a world where users control how others appear to them. In such a world, it may be possible for each person in the relationship to see their lover in very different ways than the lover might themselves choose to be seen. Is there anything wrong with using AR to view your lover as someone (or ... something ...) else? Although some philosophers have argued that imagining someone looking a certain way is ethically different from using AR to see them in that same way (Öhman 2020), it's not clear that this intuition is widely shared or that it would directly address the problems that AR might introduce into relationships.[30]

Notice that a slightly different problem can occur if users control how they appear to others. Is there anything wrong, for example, with a user choosing to represent themselves as looking, feeling, moving, and sounding in a way that differs dramatically from how they look, feel, move, and sound in unaugmented reality? Some philosophers have argued that the concept of romantic love is grounded, at least partially, on physical/sexual attraction to one's beloved (Solomon 2006; de Munck, Kronenfeld, & Manoharan 2021). If so, then it's not clear whether falling in love with a person who represents themselves in AR as appearing very differently than they would outside of AR spaces is an extension of our existing concepts of love and sex or a form of unethical misrepresentation (a phenomenon known as "catfishing") (Mosely et al. 2020). In part these issues will turn on how ongoing debates about the relationship between our unaugmented physical bodies and our sense of self evolve. Our concept of fidelity may also need to be reinterpreted to make sense of relationships in a world like this and, in the end, it may be impossible to remain completely agnostic in debates about the metaphysics of identity. Just what is it that we are being faithful to in a relationship? Are our lovers overlapping sequences of bodily states, psychological states, both, or neither (Parfit 1986)? Are they time slices of an organism (Olsen 2007)? Perhaps they're really an ongoing auto-biographical narrative that we're a part of (Schechtman 1996)? Fidelity, polyamory, monogamy, and even concepts like chastity may have to be reworked to make more sense within a world of AR embodiment, depending on how these other more fundamental questions are answered.

We can pose similar questions for friendships, though in this case the issues may be easier to address. Shannon Vallor, drawing from Aristotelian accounts of friendship, has argued that online friendships can be just as real as physical friendships under certain conditions (where users engage in empathic reciprocal interactions primarily focused on one another's well-being) (Vallor 2012). Because friendship isn't tied as strongly to physical embodiment, it seems easier to accept friendships surviving without too much trouble in a world of omnipresent AR embodiment.[31] We would still need a way of keeping track of our friends in those circumstances so that we know with whom we've engaged in reciprocal exchanges (that is, we need a solution to the second issue introduced in this chapter) but issues of embodiment pose less of a problem in friendships than they seem to in romantic relationships.

Conformity

XR worlds understandably can be viewed as spaces that greatly expand our freedom to represent ourselves. By loosening, or outright breaking, the ties that bind identity to physical bodies, XR spaces allow users to exactly tailor all aspects of their virtual bodies (how they look, feel, sound, and move). While it may seem as if a world of omnipresent AR is a world of boundless freedom for bodily expression, in this section I consider the possibility that the *opposite* may turn out to be true. In part this will turn on matters we have already looked at in this chapter. For example, we've already seen that anonymity can breed trolling. This was one (but not the only) reason why we'll need a way of keeping track of identity across AR interactions. Trolling is a kind of intentional deviance. If anonymity encourages trolling, does the lack of anonymity encourage the opposite of deviance (i.e., conformity)? There is some evidence that it does. In a study that explored whether a user's relative degree of anonymity impacted their use of language perceived by other users as offensive, Daegon Cho and Alessandro Acquisti examined a database of over 75,000 user comments on websites that varied widely on how anonymous a user's identity was. They discovered that

> [w]hen commenters use an SNS [social networking site] account (which consists in a more identifiable condition) rather than a non-SNS account, they are less likely to use offensive words. We also find that the use of real name-SNS accounts (which provide an even more identifiable condition due to the disclosure of one's real name), is associated with lower occurrence of offensive words than the case in which commenters use a pseudonym-SNS account for commenting. While the disclosure of true identity is likely to reduce the probability of using offensive words,

the greater number of users seems to prefer participating in the commenting activity by using their pseudonym accounts.

(Cho & Acquisti 2013)

Interestingly, other researchers have found that in non-anonymous social
media environments trolling behavior "does not seem to lead to a group's
decay, but rather promotes a 'natural purification' within the group through
elimination of negative influences" (Ditrich & Sassenberg 2017). How might
these findings impact user choices for XR embodiment? What would this
"natural purification" mean for AR embodiment? Assuming that we collec-
tively agree to a solution for tracking user identity over time in AR spaces, it
may be that the pressure to conform to norms about appearance may reduce
diversity in representation instead of increase it.

Contemporary movements aimed at body positivity, self acceptance, and
inclusivity are often at least partially grounded on reasons stemming from
social justice and moral luck (Lazuka et al. 2020). If ought implies can (i.e.,
if moral obligations only apply to someone if they're possible for that
person to meet), then it may be unfair to require people to exert tremendous
efforts to change how their bodies look or to whom they are attracted
(Bennett 2014). These arguments are meant to lead us to the conclusion that
it's better for us to re-examine the norms that impose burdens on others to
change how they look or feel than to keep holding people to these unfair
expectations. This is especially true when we're dealing with norms calling
for users to change features that are impossible or extremely difficult to
change, or if they're about essentially contestable topics (i.e., debates about
which people could reasonably disagree with one another) (Ramirez 2017).
A world of omnipresent AR embodiment may disrupt self-acceptance
movements in unexpected ways.

Philosopher Michael Sandel, discussing the ethics of genetic engineering,
provides some helpful arguments for us. A world of finely-tuned genetic
engineering resembles, in many ways, a world of omnipresent AR where
users have radical freedom in terms of how they're embodied. As one user
put it in a recent study of avatar customization using VR: "in real life,
you're stuck with what you were born with. But in VR, you can be what
you truly feel like you are inside" (quoted in Freeman et al. 2020). Sandel
was worried that, in a world where genetic engineering would give everyone
"mastery" over their bodies, a hidden problem would materialize:

The deeper danger [of worlds where users exercise control over all
aspects of their bodies] is that they represent a kind of hyperagency – a
Promethean aspiration to remake nature, including human nature, to
serve our purposes and satisfy our desires. The problem is not the drift
to mechanism but the drive to mastery. And what the drive to mastery
misses and may even destroy is an appreciation of the gifted character
of human powers and achievements. To acknowledge the giftedness of

life is to recognize that our talents and powers are not wholly our own doing, despite the effort we expend to develop and exercise them. It is also to recognize that not everything in the world is open to whatever use we may desire or devise.

(Sandel 2007)

In part, Sandel argues that humans everywhere have all faced a similar challenge of taking their bodies and brains as they were given to them and then confronting the challenge of making the best life possible given the limitations imposed upon them by that biology. This breeds in us, he argued, a kind of humility and also a sense of empathy with other people (who are all making do as best they can too). Sandel was worried that

[a]s humility gives way, responsibility expands to daunting proportions. We attribute less to chance and more to choice. Parents become responsible for choosing, or failing to choose, the right traits for their children. Athletes become responsible for acquiring, or failing to acquire, the talents that will help their teams win ... today when a basketball player misses a rebound, his coach can blame him for being out of position. Tomorrow the coach may blame him for being too short.

(Sandel 2007)

We can imagine the same consequences for our AR bodies. A world where bodies are not only malleable but *easily* and *instantly* malleable is a world where body acceptance may mean conforming to pre-existing norms about beauty, ability, and so on. It may mean a world where our worst norms about race, gender, nationality, and language are conformed to out of ease instead of challenged as unethical or oppressive. This is not a foregone conclusion but a world where our bodies, movements, and even the language we're heard to speak (and *how* it's spoken) can be easily tailored or changed is a world where diversity is arguably less likely to flourish. This may even be a reason that speaks against tracking user identity over time (at the very least it should be a part of the global risk assessment carried by these technologies as they become integrated into the basic structure of society).

Conclusion

This has been a book exploring the psychology and ethics of building worlds. I've argued that we need to take our virtual experiences much more seriously than we currently do. AR and VR are poised to change the nature of human experience and our social structures at least as radically as the internet has. At their best these technologies make us better scientists (by shifting us away from the Thought Experiment Paradigm), help us to be better people (by expanding our learning opportunities and cultivating sympathy), and even let us do impossible things like visit Mars.[32] At their worst, AR and VR are capable of helping people inflict very real trauma on others, of bringing out the worst aspects of our personalities, and can even undermine hard fought battles like the push for an intersectional approach to identity, oppression, and experience.

In this chapter I looked further into the future than in the rest of the book. A world of omnipresent XR spaces is a world our preaugmented concepts are not ready to deal with. The heuristics we've evolved to keep track of identity, our concepts of love, fidelity, and friendship, and even arguments about the value of diversity and acceptance need to be ready (in some cases revised) to confront the forthcoming XR landscape. The arguments in this chapter are, more than those in the rest of the book, conjectural and subject to revision. Unforeseen technological and social developments are likely to change the ultimate shape of our AR future but it's not likely to change the fact that these technologies are here to stay and that they'll alter the ethical landscapes around us. We need to be ready for these changes lest we repeat the technological missteps of the 20th century.

Notes

1 I'll expand on these heuristics later in the chapter. As one example, individuals with a condition known as prosopagnosia have a much harder time than neurotypical people when it comes to recognizing other people's faces (Kress & Daum 2003). I might have to develop other skills and heuristics to recognize my friend at the coffee shop if I wasn't able to rely on help from my fusiform facial gyrus to handle this for me (Kanwisher, McDermott, & Chun 1997).

2 We'll have good reasons to question whether we *want* our concepts of identity to be stable and sticky (both Western and Non-Western theorists about identity have resisted the idea that our identities work this way). At the very least, there will be at least some occasions (paying taxes, voting, etc.) where stability and stickiness are desirable.

3 I'll say much more later in this chapter about contemporary debates regarding the nature of various forms of trans identities and what these discussions should tell us about contemporary views regarding the relationship between the body, the self, and representation. In short, although contemporary debates about the nature of race, gender, and able-bodiedness are complex, they point in the direction of severing the connection between identity and the body (though, of course, not without some critics).

4 Although not currently possible, we can imagine AR overlays that include sensory models beyond sight and sound. In Chapter 2 I discussed the importance of haptics as a component of both presence and virtually real experiences. We can imagine haptic AR overlays as a component of the XR spectrum as well that would both change how we experience existing objects but also allow us to experience entirely simulated objects in more realistic ways (Stone 2000).

5 Members of the "Nullo" community, for example, have specific parts of their bodies surgically removed as a form of self-expression (Nullo 2021).

6 This is perhaps most clearly seen in contemporary debates over the nature of trans identities other than transgender identity. For example, some individuals,

self-identifying as having a condition referred to as Body Integrity Identity Disorder, claim a trans-abled identity where their external presentation and internal sense of self do not match. Individuals identifying in this way sometimes request, to some controversy, that they receive surgical treatments intended to amputate otherwise healthy limbs or organs in order to help their external and internal senses of self come into alignment (Müller 2009). Similarly, individuals who claim a controversial trans-racial identity (Tuvel 2017) will often seek cosmetic interventions of different kinds in order to match their external physical presentation with an internal sense of their racial identity. I mention these instances here, as I do later in this chapter, not intending to stake a position on the ethical and metaphysical debates raised over the status of these identities but, instead, to note that the ethics of representation (representational intersectionality) looms large over all of these debates.

7   In that sense, instances of body dysmorphic disorder have been rising among men, although the way that BDD presents in men is a result of differing gender norms. Men with BDD tend to focus on their perceived lack of musculature instead of perceived weight or fat (Pope et al. 2005).

8   This seems like a pessimistic way of addressing issues about toxic beauty norms. Instead of helping individuals become more comfortable with their bodies "as they are" and instead of educating individuals about the harmful sources of their internalized beauty standards, this solution seems to give in completely to such standards. Although this is true (and although I agree that it would be ideally better to address the toxicity of norms), it seems more likely to me that, faced with the question of weather to overhaul beauty norms or to use technology to empower people to modify their bodies as they see fit that, as a descriptive social psychological prediction, we're more likely to choose the latter option than the former.

9   For example, the code offers guidance on content moderation in VR and AR, on the special case of children being exposed to VR and AR simulations, and so on.

10 In a very real sense, either of the first two responses to questions about control and representation will presume a commitment to the third or fourth option as well. For example, if we conclude that individual users should be empowered to embody themselves, and to be seen by others however they wish in AR spaces, then we're simultaneously calling for some kind of regulatory (if state) or policy (if corporate) framework be put in place to support that position.

11 The revision to the state law made it such that "any person who photographs or records by any means the image of the intimate body part or parts of another identifiable person, under circumstances where the parties agree or understand that the image shall remain private, and the person subsequently distributes the image taken, with the intent to cause serious emotional distress, and the depicted person suffers serious emotional distress, is guilty of disorderly conduct and subject to that same punishment" (SB255 2013)

12 The Philippines outlawed the non-consensual distribution of sexually explicit images in 2009.

13 Our concepts of race, gender, identity, and the body are all currently concepts that are subject to active debate and disagreement. I don't intend to step into those debates here and am not sure how participants in such a debate would view someone whose physical body presented as White but who genuinely identified as Black, for example, choosing to adopt a Black-presenting AR body (Barnes 2019; Bettcher 2013; Bogardus 2020; Butler 1999; Haslanger 2000; Tuvel 2017). As I'll say later in this chapter, what's important for our purposes is that our sense of the relationship between body and identity has changed significantly in the last 30 years and it's likely to continue to change as the moral, social, and political debates about trans identities of all sorts settle into equilibrium.

14 This is of course assuming one does not live in countries that block access to large swathes of the internet, like mainland China (Ensafi et al. 2015), Libya, Sudan, Kuwait, Bahrain, Oman, Mauritania, Qatar, or Jordan (Shishkina & Issaev 2018).

15 A related issue here was introduced by philosopher Carl Öhman. Öhman introduces what he calls "The Pervert's Dilemma" as a way of bringing to light the seeming contradiction between the moral permissibility of fantasizing about someone being nude as opposed to using technology to digitally create a nude image of the same person. "To be clear, the contradiction of the pervert's dilemma is thus not that sexual fantasies never can be impermissible, while deepfakes are always are impermissible, but rather that a representation that would (normally) be deemed permissible as a fantasy is deemed impermissible as a deepfake, despite the absence of any immediately identifiable and morally relevant distinction between the two formats. Thus, given that one accepts 1 and 2, it seems that one must either accept that deepfake content is morally acceptable as long as conditions (i) and (ii) are fulfilled or accept that sexual fantasies are morally objectionable despite not directly harming anyone. Neither option seems intuitively right" (Öhman 2020).

16 Some readers may get to this point and not see a problem with this use of AR technology. Along consequentialist lines they may think a world with less overt racism and bias is a better world, all things considered. I think that there's something to this argument and don't mean to discount this as a knock-down argument against empowering users to control how others appear to them. Whatever one's thoughts about this use of AR, it must be taken into account in any assessment of the ethics of omnipresent AR spaces.

17 Although Neely is focused on AR overlays of property and not AR embodiment, she comes to a similar conclusion as I do here about the relationship between the ethics of AR overlays and state regulation: "if we see augmentation as graffiti – albeit frequently helpful or interesting graffiti – then it seems more within the state's purview to control. With the exceptions of a few artists like Banksy, graffiti is seen as a nuisance, not something which benefits the public, and we do allow its erasure. In this case, augmentation would be something that could be prevented and/or removed as the state sees fit; while we protect speech in public places (in general), we do not protect all forms of action – there are things you are not permitted to do in a national park, and augmentation might simply be added to the list. Part of the problem is that augmentations will likely be some of each of these. A person who goes about augmenting public places with photos of his genitals is essentially engaging in the visual equivalent of spam – this sort of augmentation seems akin to graffiti. A person who augments a place with historical information seems more like speech. As for a person who augments a museum to tell you that their café serves terrible coffee, well, whether it is more like speech or graffiti in that case seems to be in the eye of the beholder" (Neely 2019). In this sense, although speech may be something users have the freedom to engage or not to engage in, states need a regulatory framework for AR speech that occurs on AR platforms. Like Floridi, it may be that Neely is treating AR spaces like these as an AR commons.

18 Maguinness and Newell go on to explore how facial processing changes with age. Their results are fascinating: "This experience-dependent malleability suggests that face recognition is also a product of ontogenetic adaptation (i.e., individual development across the lifespan), which has direct consequences on our ability to discriminate and remember faces" over time (2014).

19 Whether or not the FFG is an evolved module is a matter of serious debate among neuroscientists. Some, for example, argue that the FFG evolved to help us engage in all fine-grained perceptual tasks that require reidentification (Haist,

Lee, & Stiles 2010) and thus that this neurological structure didn't evolve only, or even uniquely, to detect faces. I think we can remain agnostic about the evolutionary origins of the so-called FFG here. What matters is that, regardless of whether we evolved a module to do this, that facial information is one of our most natural means of reidentification.

20 Olfaction is a complex sense. In this study, Manfred Milinski and colleagues came to the conclusion that their "volunteers recognized the supplementation of their body odour by MHC peptides and preferred 'self' to 'non-self' ligands when asked to decide whether the modified odour smelled 'like themselves' or 'like their favourite perfume'. Functional magnetic resonance imaging indicated that 'self'-peptides specifically activated a region in the right middle frontal cortex. Our results suggest that despite the absence of a vomeronasal organ, humans have the ability to detect and evaluate MHC peptides in body odour" (Milinski et al. 2013). As with the FFG, any claim about neurological modules should be read with caution. Having said that, Milinski et al.'s findings don't depend on the existence of such a module. Their subjects were able to recognize their own odor whether or not they were using a distinctly evolved module to do it.

21 Readers interested in learning more about the epistemological question are welcome to explore possible answers outlined in the *Stanford Encyclopedia of Philosophy* and its entry on "Personal identity" (Olsen 2019). Readers interested in exploring questions about the metaphysical status of virtual objects are highly encouraged to check out David Chalmers' work on the subject (Chalmers 2017).

22 Notice that we can track faces, sounds, smells, and movements even if we end up concluding that enduring selves are an illusion. These evolutionary capacities are agnostic (or, at least, compatible with a wide variety of answers) to metaphysical responses about personal identity.

23 Metaphysically this can get tricky. For example, John Locke (1975/1694) famously posed a thought experiment wherein two people switched psychologies (e.g., Person A's memories are inserted into Person B's body and vice versa). He did this as a way to ask readers just how much the body itself matters to our concept of identity. For Locke, metaphysically, we go where our psychology goes and not where our body happens to be. Luckily for us, we're avoiding these metaphysical questions. Whether or not we are, deep down, our bodies, psychologies, organisms, or some combination of views is a question I'd prefer to sidestep here in favor of the more practical question: regardless of what we are, how can we track identity for the purposes of things like voting, shopping, and personal relationships?

24 Does this remove our freedom to modify our bodies? In a world of *omnipresent AR* I think the answer is yes. If we're always, or nearly always, in an AR space then our actual physical appearance plays a much less significant role in how we're thought about and viewed. Reducing this to a kind of costume that hides their "real self" trivializes the degree to which people can (and probably will) identify themselves with their AR body more than their physical body. Snapchat and Zoom dysphorias already show us that at least some people already identify more strongly with their AR self than their physical self.

25 Although "Gender Dysphoria" (GD) has, in most ways, replaced "Gender Identity Disorder" (GID), the differences between those disorder categories is important. GID sees one's subjective sense of gender identity as dysfunctional and in need of treatment. For GD the dysfunction is not identified with one's sense of gender identity but instead, as the name implies, with the dysphoria (unhappiness/suffering) from a mismatch between one's sense of gender identity and one's sex assigned at birth. Although this is a welcome improvement in the diagnostic language, many trans advocates are calling for the removal of GD from the DSM not only on metaphysical grounds (that unhappiness caused by one's gender

identity is a social problem not a mental illness) but also on ethical ones (that the stigma of mental illness hurts trans persons more than it helps) (Lev 2013).

26 For some readers this may not be a cost at all (Stock 2019). For such readers, the biomarker / embodiment approach may be more attractive. Having said that, even for readers with these sorts of views, the biomarker approach to fixing identity would still need to confront issues related to privacy and authoritarianism, and cannot fully evade questions about the body in an AR space. For example, gender critical feminists like Kathleen Stock (cited above) argue that "female" is a biological category and hence cannot be changed in the same way that gender (a socially constructed category) can be. This is because they view misogyny as resting (largely) on biological oppression. In AR spaces, biology becomes significantly less visible (a female, on gender critical views, can very easily appear as male or as non-sexed in AR spaces). We would still need, in other words, some way of keeping track of identity even if we think that contemporary debates about sex and the body should be settled by biology.

27 We'll note that privacy of this kind is probably impossible to hold on to in a world of omnipresent AR overlays where we solve the tracking problem we're looking at now. It might be said that this isn't that much of a loss as users appear unfazed today at the prospect of giving corporations like Alphabet (the parent company of Google) constant real time data about where they are, where they're going, and so on by using their Maps software (Ketelaar & Balen 2018).

28 Those who are curious about the work can view it at the artist's website: https://www.beeple-crap.com/everydays

29 It is estimated that as many as 20 percent of all Bitcoins (worth an estimated 140 billion USD as of 2021) are "lost" (i.e., their owners have lost the key that allows them to access them and hence are permanently unavailable for trading) (Lost bitcoin 2021).

30 Carl Öhman, or a nearby critic, might reply that in many ways romantic partners regularly see one another differently. Robert Solomon, for example, has argued that idealization of one's beloved is a necessary condition for romantic love (Solomon 2006). If that's the case then perhaps Öhman's (2020) dilemma can be resurrected: if lovers already see one another very differently from how they see themselves, is there a morally relevant difference between this de facto state of affairs and lovers using AR overlays to do this work for them?

31 Vallor herself may not be fully committed to this claim. For example, in the same article she claims that self-knowledge is an essential element of empathy and that empathy is necessary for virtuous friendships. In describing the process of self-knowledge, Vallor notes that "I build up an image of a self-contained being that I am: made up of a certain kind of body, mind and/or soul. Self-knowledge is not, for Aristotle, a matter of 'going inside' to observe some private, autonomous and unique inner core of the personality, as we often portray it in the modern West. Instead, self-knowledge in the Aristotelian sense is a matter of understanding properly where I fit in the world, what my proper role in it is, and the capacities I have (or lack) for actively flourishing in those roles" (Vallor 2012). If self-knowledge requires knowledge of one's *physical* unaugmented body then self-knowledge cuts against thinking of one's AR body as a part of the self. It's also possible that self-knowledge in AR spaces requires knowledge only of one's AR embodiment (in the same way, we might say, that self-knowledge need not require an understanding of the physical forces that constitute the atoms that make up the cells of our physical bodies).

32 If, by the time you're reading this, Martian tourism is a reality then replace Martian tourism with exoplanet tourism!

References

ACM (Association for Computing Machinery). 2018. ACM code of ethics and professional conduct. Retrieved from https://www.acm.org/code-of-ethics

American Psychiatric Association. 2000. *Diagnostic and statistical manual of mental disorders: DSM-IV-TR*. Washington DC: APA.

American Psychiatric Association. 2013. *Diagnostic and statistical manual of mental disorders: 5th ed*. Washington DC: APA.

Barnes, E. 2019. Gender and gender terms. *Noûs*, 54 (3), 704–730.

Bayne, T., & Levy, N. 2005. Amputees by choice: Body Integrity Identity Disorder and the ethics of amputation. *Journal of Applied Philosophy*, 22 (1), 75–86.

Bennett, J. 2014. "Born this way": Queer vernacular and the politics of origins. *Communication and Critical/Cultural Studies*, 11 (3), 211–230.

Bettcher, T.M. 2013. Trans women and the meaning of "woman." In Soble, A., Power, N., & Halwani, R. (Eds.), *Philosophy of sex: Contemporary readings*. 6th ed. Lanham MD: Rowman & Littlefield, 233–250.

Bogardus, T. 2020. Evaluating arguments for the sex/gender distinction. *Philosophia*, 48, 873–892.

Brooker, C. (Writer), & Tibbetts, C. (Director). 2014, December 16. White Christmas. In Jones, A. & Brooker, C. (Executive producers), *Black Mirror*. House of Tomorrow.

Butler, J. 1999. *Gender trouble*. London: Routledge.

ByteDance. 2019. TikTok. Retrieved from https://www.tiktok.com

Chalmers, D. 2017. The virtual and the real. *Disputatio*, 9 (46), 309–352.

Cho, D., & Acquisti, A. 2013. The more social cues, the less trolling? An empirical study of online commenting behavior. *Semantic Scholar*. Carnegie Mellon University. https://doi.org/10.1184/R1/6472058.v1

Cillizza, C. 2021, May 5. Trump is banned from Facebook. Trumpism is everywhere on it. CNN. Retrieved from https://edition.cnn.com/2021/05/05/politics/donald-trump-facebook/index.html

Cooper, A. 1999. Sexuality and the Internet: Surfing into the new millennium. *CyberPsychology and Behavior*, 1, 181–187.

Cornwell, B., & Lundgren, D. 2001. Love on the Internet: Involvement and misrepresentation in romantic relationships in cyberspace vs. realspace. *Computers in Human Behavior*, 17, 197–211.

Crenshaw, K. 1991. Mapping the margins: Intersectionality, identity politics, and violence against women of color. *Stanford Law Review*, 43 (6), 1241–1299.

Culver, C.M., & Gert, B. 1990. The inadequacy of incompetence. *Milbank Quarterly*, 68 (4) 619–643.

Culver, C.M., & Gert, B. 2004. Competence. In *The philosophy of psychiatry: A companion*. Oxford: Oxford University Press, 258–271.

Cutting, J.E., & Kozlowski, L.T. 1977. Recognizing friends by their walk: Gait perception without familiarity cues. *Bulletin of the Psychonomic Society*, 9, 353–356.

Danesi, M. 2018. Makeup: Why do we put it on? In *Of cigarettes, high heels, and other interesting things*. New York: Palgrave Macmillan. https://doi.org/10.1057/978-1-349-95348-6_3

de Munck, V.C., Kronenfeld, D.B., & Manoharan, C. 2021. A prototype analysis of the cultural and evolutionary construction of romantc love as a synthesis of love and sex. *Journal of Cognition and Culture*, 21(1–2), 25–48.

de Ruiter, A. 2021. The distinct wrong of deepfakes. *Philosophy and Technology*, https://doi.org/10.1007/s13347-021-00459-2

Ditrich, L., & Sassenberg, K. 2017. Kicking out the trolls: Antecedents of social exclusion intentions in Facebook groups. *Computers in Human Behavior*, 75, 32–41.

Dowling, M. 2021. Fertile LAND: Pricing non-fungible tokens. *Finance Research Letters*. https://doi.org/10.1016/j.frl.2021.102096

Ensafi, R., Winter, P., Mueen, A., & Crandall, J.R. 2015. Analyzing the Great Firewall of China over space and time. *Proceedings on Privacy Enhancing Technologies*, 1, 61–76.

Fallahi, H.R., Keyhan, S.O., Cheshmi, B., Zandian, D., & Moghadam, P.J. 2021. Augmented reality: New horizons in oral and maxillofacial surgery. In Keyhan, S. O., Fattahi, T., Bagheri, S.C., Bohluli, B., & Amirzade-Iranaq, M.H. (Eds.), *Integrated procedures in facial cosmetic surgery*. Cham, Switzerland: Springer. https://doi.org/10.1007/978-3-030-46993-1_51

Fernandes Botts, T. 2018. In black and white: A hermeneutic argument against "transracialism." *Res Philosophica*, 95 (2), 303–329.

Finnegan, D.J., Zoumpoulaki, A., & Eslambolchilar, P. 2021. Does mixed reality have a Cassandra complex? *Frontiers in Virtual Reality*. doi:10.3389/frvir.2021.673547

Floridi, L. 2021. Trump, Parler, and regulating the infosphere as our commons. *Philosophy and Technology*, 34, 1–5.

Freeman, G., Zamanifard, S., Maloney, D., & Adkins, A. 2020. My body, my avatar: How people perceive their avatars in social virtual reality. In *CHI EA '20: Extended abstracts of the 2020 CHI conference on Human Factors in Computing Systems*, 1–8.

Fung, B. 2021, January 9. Twitter bans President Trump permanently. CNN. Retrieved from https://edition.cnn.com/2021/01/08/tech/trump-twitter-ban/index.html

Gillon, R. 2015. Defending the four principles approach as a good basis for good medical practice and therefore for good medical ethics. *Journal of Medical Ethics*, 41, 111–116.

Golf-Papez, M., & Veer, E. 2017. Don't feed the trolling: Rethinking how online trolling is being defined and combated. *Journal of Marketing Management*, 33 (15–16), 1336–1354.

Guo, C., & Caine, K. 2021. Anonymity, user engagement, quality, and trolling on Q&A sites. *Proceedings of the ACM on Human-Computer Interaction*, 5 *(CSCW1)*, Article no. 141. https://doi.org/10.1145/3449215

Haist, F., Lee, K., & Stiles, J. 2010. Individuating faces and common objects produces equal responses in putative face-processing areas in the ventral occipito-temporal cortex. *Frontiers in Human Neuroscience*, 8 (4), 181. doi:10.3389/fnhum.2010.00181.

Hansson, E., Elander, A., Hallberg, H., & Sandman, L. 2020. Should immediate breast reconstruction be performed in the setting of radiotherapy? An ethical analysis. *Journal of Plastic Surgery and Hand Surgery*, 54 (2), 83–88. doi:10.1080/2000656X.2019.1688165.

Hargie, O.D.W., Mitchell, D.H., & Somerville, I.J.A. 2015. "People have a knack of making you feel excluded if they catch on to your difference": Transgender experiences of exclusion in sport. *International Review for the Sociology of Sport*, 52 (2), 223–239.

Haslanger, S. 2000. Gender and race: (What) are they? (What) do we want them to be? *Noûs*, 34 (1), 31–55.

Higgins, S., & Wysong, A. 2018. Cosmetic surgery and Body Dysmorphic Disorder: An update. *International Journal of Women's Dermatology*, 4 (1), 43–48.

Hilton, E.N., & Lundberg, T.R. 2021. Transgender women in the Female category of sport: Perspectives on testosteron suppression and performance advantage. *Sports Medicine*, 51, 199–214.

Johnson, D.G., & Diakopoulos, N. 2021. What to do about deepfakes. *Communications of the ACM*, 64 (3), 33–35.

Kanwisher, K., McDermott, J., & Chun, M.M. 1997. The Fusiform Face Area: A module in human extrastriate cortex specialized for face perception. *Journal of Neuroscience*, 17 (11), 4302–4311.

Kaplan, J.T., Aziz-Zadeh, L., Uddin, L.Q., & Iacoboni, M. 2008. The self across the senses: an fMRI study of self-face and self-voice recognition. *Social Cognitive and Affective Neuroscience*, 3 (3), 218–223. https://doi.org/10.1093/scan/nsn014

Ketelaar, P., & Balen, M. 2018. The smartphone as your follower: The role of smartphone literacy in the relation between privacy concerns, attitude and behaviour towards phone-embedded tracking. *Computers in Human Behavior*, 78, 174–182.

Kress, T., & Daum, R. 2003. Developmental prosopagnosia: A review. *Behavioural Neurology*, 14, 109–121.

Lazuka, R.F., Wick, M.R., Keel, P.K., & Harriger, J.A. 2020. Are we there yet? Progress in depicting diverse images of beauty in Instagram's body positivity movement. *Body Image*, 34, 85–93.

Lev, A.I. 2013. Gender dysphoria: Two steps forward, one step back. *Clinical Social Work Journal*, 41, 288–296.

Li, X., Jiang, P., Chen, T., Luo, X., & Wen, Q. 2020. A survey on the security of blockchain systems. *Future Generation Computer Systems*, 107, 841–853.

Locke, J. 1975/1694. *An essay concerning human understanding*. Nidditch, P. (Ed.). Oxford: Clarendon Press.

Lost bitcoin. 2021, January 13. Man has two guesses to unlock bitcoin worth $240m. BBC. Retrieved from https://www.bbc.com/news/technology-55645408

Mac, Ryan. 2019, June 27. Trump Tweets that violate Twitter's rules will now get a warning label. *Buzzfeed News*. Retrieved from https://www.buzzfeednews.com/article/ryanmac/twitter-world-leader-trump-tweets-warning-label

Madary, M., & Metzinger, T.K. 2016. Real virtuality: A code of ethical conduct. Recommendations for good scientific practice and the consumers of VR-Technology. *Frontiers in Robotics and AI*. https://doi.org/10.3389/frobt.2016.00003

Maguinness, C., & Newell, F.N. 2014. Recognizing others: Adaptive changes to person recognition throughout the lifespan. In Schwartz, B.L., Howe, M.L., Toglia, M.P., & Otgaar, H. (Eds.), *What is adaptive about adaptive memory?* Oxford: Oxford University Press, 231–257.

Milinski, M., Croy, I., Hummel, T., & Boehm, T. 2013. Major histocompatibility complex peptide ligands as olfactory cues in human body odour assessment. *Proceedings of the Royal Society B*, 280, 20122889. http://dx.doi.org/10.1098/rspb.2012.2889

Modiface. 2019. Retrieved from https://modiface.com/

Mora, C., Rollins, R.L., Taladay, K., Kantar, M.B., Chock, M.K., Shimada, M., & Franklin, E.C. 2018. Bitcoin emissions alone could push global warming above 2° C. *Nature Climate Change*, 8, 931–933.

Moscufo, M. 2021. March 11. Digital artwork sells for record $69 million at Christie's first NFT auction. NBC News. Retrieved from https://www.nbcnews.com/business/business-news/digital-artwork-sells-record-60-m illion-christie-s-first-nft-n1260544

Mosley, M.A., Lancaster, M., Parker, M.L., & Campbell, K. 2020. Adult attachment and online dating deception: A theory modernized. *Sexual and Relationship Therapy*, 35, 227–243.

Müller, S. 2009. Body integrity identity disorder (BIID) – Is the amputation of healthy limbs ethically justified? *American Journal of Bioethics*, 9 (1), 36–43.

Munn, N.J. 2012. The reality of friendship within immersive virtual worlds. *Ethics and Information Technology*, 14, 1–10.

Neely, E.L. 2019. Augmented reality, augmented ethics: who has the right to augment a particular physical space? *Ethics and Information Technology*, 21, 11–18.

Nelson, L., & Ramirez, E. 2017. Can suicide in the elderly be rational? In McCue, R. E. & Balasubramania, M. (Eds.), *Rational Suicide in the Elderly: Clinical, Ethical and Sociocultural Aspects*. New York: Springer, 1–21.

Nullo (body modification). 2021, June 5. *Wikipedia*. Retrieved from https://en.wikip edia.org/wiki/Nullo_(body_modification)

Öhman, C. 2020. Introducing the pervert's dilemma: A contribution to the critique of deepfake pornography. *Ethics and Information Technology*, 22, 133–140.

Olsen, E.T. 2007. *What are we? A study in personal ontology*. Oxford: Oxford University Press.

Olsen, E.T. 2019, September 6. Personal identity. In *Stanford Encyclopedia of Philosophy*. Retrieved from https://plato.stanford.edu/entries/identity-personal/

Overall, C. 2004. Transsexualism and "transracialism". *Social Philosophy Today*, 20, 183–193.

Parfit, D. 1971. Personal identity. *The Philosophical Review*, 80 (1), 3–27.

Parfit, D. 1986. *Reasons and persons*. Oxford: Oxford University Press.

Pase, S. 2012. *Ethical considerations in augmented reality applications*. 2012 EEE International Conference on e-learning, e-business, Enterprise Information Systems, and e-government, Las Vegas NV.

Pertaub, D.P., Slater, M., & Barker, C. 2002. An Experiment on public speaking anxiety in response to three different types of virtual audience. *Presence Teleoperators and Virtual Environments*, 11 (1), 68–78.

Pope, C.G., Pope., H.G., Menard, W., Fay, C., Olivardia, R., & Philips, K.A. 2005. Clinical features of muscle dysmorphia among males with body dysmorphic disorder. *Body Image*, 2 (4), 395–400.

Ramirez, E. 2017. A conditional defense of shame and shame punishment. *Symposion: Theoretical and Applied Inquiries in Philosophy and Social Sciences*, 4 (1), 77–95.

Ramphul, K., & Mejias, S.G. 2018, March 3. Is "Snapchat Dysmorphia" a real issue? *Cureus* 10 (3), e2263. doi:10.7759/cureus.2263.

Rice, S.M., Graber, E., & Kourosh, A.S. 2020. A pandemic of dysmorphia: "Zooming" into the perception of our appearance. *Facial Plastic Surgery and Aesthetic Medicine*, 22 (6), 401–402.

Rudd, A. 2015. No self?: Some reflections on Buddhist theories of personal identity. *Philosophy East and West*, 65 (3), 869–891.

Sandel, M.J. 2007. *The case against perfection: Ethics in the age of genetic engineering*. Cambridge MA: Harvard University Press.

SB 255 Disorderly conduct: invasion of privacy. 2013. California Senate. Retrieved from https://leginfo.legislature.ca.gov/faces/billNavClient.xhtml?bill_id=201320140SB255

Schechtman, M. 1996. *The constitution of selves.* Ithaca NY: Cornell University Press.

Schechtman, M. 2005. Personal identity and the past. *Philosophy, Psychiatry, and Psychology,* 12 (1), 9–22.

Schmeiss, J., Hoelzle, K., & Tech, R.P.G. 2019. Designing governance mechanisms in platform ecosystems: Addressing the paradox of openness through blockchain technology. *California Management Review,* 62 (1), 121–143.

Sealey, K. 2018. Transracialism and White allyship: A response to Rebecca Tuvel. *Philosophy Today,* 62 (1), 21–29.

Shishkina, A., & Issaev, L. 2018. Internet censorship in Arab countries: Religious and moral aspects. *Religions,* 9, 358.

Siltanen, S., Oksman, V., & Ainasoja, M. 2013. User-centered design of augmented reality interior design service. *International Journal of Arts and Sciences,* 6 (1), 547–563.

Snap Inc. 2011. Retrieved from https://www.snapchat.com/add/citeapprenti.e

Solomon, R. 2006. *About love: Reinventing romance for our times.* Indianapolis: Hackett.

Steele, P., Burleigh, M.K., Myrene, M., & Bailey, L. 2020. Ethical considerations in designing virtual and augmented reality products – virtual and augmented reality design with students in mind: Designer's perceptions. *Journal of Educational Technology,* 49 (2), 219–238.

Stock, K. 2019. Sexual orientation: What is it? *Proceedings of the Aristotelian Society,* 119 (3), 295–319.

Stone, R.J. 2000. Haptic feedback: A brief history from telepresence to virtual reality. In Brewster, S. & Murray-Smith, R. (Eds.), *Haptic Human–Computer Interaction.* New York: Springer, 1–16.

Su, J., Shukla, A., Goel, S., & Narayanan, A. 2017. De-anonymizing web browsing data with social networks. In *WWW '17: Proceedings of the 26th International Conference on World Wide Web,* 1261–1269.

Sugiura, M., Watanabe, J., Maeda, Y., Matsue, Y., Fukuda, H., & Kawashima, R. 2005. Cortical mechanisms of visual self-recognition. *NeuroImage,* 24 (1), 143–149.

Sugunasiri, S.H.J. 2011. "Asoulity" as translation of anattā: Absence, not negation. *Canadian Journal of Buddhist Studies,* 7, 101–134.

Tiku, N., & Newton, C. 2015, February 4. Twitter CEO: "We suck at dealing with abuse." *The Verge.* Retrieved from https://www.theverge.com/2015/2/4/7982099/twitter-ceo-sent-memo-taking-personal-responsibility-for-the

Truby, J. 2018. Decarbonizing Bitcoin: Law and policy choices for reducing the energy consumption of Blockchain technologies and digital currencies. *Energy Research and Social Science,* 44, 399–410.

Tuvel, R. 2017. In defense of transracialism. *Hypatia,* 32 (2), 263–278.

Vallor, S. 2012. Flourishing on facebook: Virtue friendship and new social media. *Ethics and Information Technology,* 14, 185–199.

Vallor, S. 2015. Moral deskilling and upskilling in a new machine age: Reflections on the ambiguous future of character. *Philosophy and Technology,* 28 (1), 107–124.

Westerlund, M. 2019. The emergence of deepfake technology: A review. *Technology Innovation Management Review,* 9 (11), 40–53. http://doi.org/10.22215/timreview/1282

Wolf. 2021, June 5. Naomi Wolf banned from Twitter for spreading vaccine myths. *The Guardian*. Retrieved from https://www.theguardian.com/books/2021/jun/05/naomi-wolf-banned-twitter-spreading-vaccine-myths

Young, R. 2021, April 16. A LeBron James Top Shot moment that honored Kobe Bryant sold for nearly $400,000. *Yahoo! Sports*. Retrieved from https://sports.yahoo.com/nba-lebron-james-top-shot-moment-nft-honoring-kobe-bryant-dunk-helicopter-crash-sold-auction-173649230.html.

Index